Introduction

The aim of the Primary Mathematics curriculum is to allow students to develop their ability in mathematical problem solving. This includes using and applying mathematics in practical, real-life situations as well as within the discipline of mathematics itself. Therefore, the curriculum covers a wide range of situations from routine problems to problems in unfamiliar contexts to open-ended investigations that make use of relevant mathematical concepts.

An important feature of learning mathematics with this curriculum is the use of a concrete introduction to the concept, followed by a pictorial representation, followed by the abstract symbols. The textbook supplies the pictorial and abstract aspects of this progression. You, as the teacher, should supply the concrete introduction. If the topic is new to the student, provide the concrete introduction before doing the tasks in the text. Use the tasks in the textbook more as an assessment of understanding. If the topic is mostly review, you can use concrete objects if you detect some misunderstanding. For some students a concrete illustration is more important than for other students.

The textbook and the workbook are a basis for developing mathematical reasoning skills. Practicing procedures or "math facts" can easily be added through additional worksheets for students who need more practice. As you go through this curriculum, it is important that you gauge your own student's understanding of concepts and need for additional practice and provide it when needed. The textbook is used for discussion, learning, and assessment by the teacher as part of the lesson. Some pages or tasks are used for discussion that can go beyond the obvious answer. Other tasks are straightforward ones where you student simply supplies the answer. The workbook is for independent practice.

The workbook contains several reviews. You can use these in any way beneficial to your student. For students who benefit from a more continuous review, you can assign 3 problems a day from one of the reviews. Or, you can use the reviews to assess any misunderstanding before administering a test. The reviews are cumulative, and so allow you to refresh your student's memory or understanding on a topic that was covered earlier in the year. In addition, there are supplemental books for Extra Practice and Tests. In the test book, there are two tests for each section. The second test is multiple choice. There is also a set of cumulative tests at the end of each unit. You do not need to use both tests. If you use only one test, you can save the other for review or practice later on. You can even use the review in the workbook as a test, and not get the test book at all. So there are plenty of choices for assessment, review, and practice.

The purpose of this guide is to help you to understand the important concepts of the *Primary Mathematics* curriculum, to gain an understanding of how these concepts fit in with the program as a whole, to provide suggestions to help you introduce concepts concretely and use the textbook effectively, and to provide suggestions for additional activities for reinforcement and practice. You should pick and choose the activities that are most useful for your particular student – you do not have to do every activity.

This guide will give you a suggested amount of time in weeks to spend on each unit. For some units, your child may be able to do the work more quickly, and for others more slowly. Take the time your student needs on each topic. For one topic you may have to add some time for mastering a concept in one topic, and another topic may be completed faster than the suggested time.

Scheme of Work

Textbook: *Primary Mathematics, Standards Edition, 1B Textbook*
Workbook: *Primary Mathematics, Standards Edition, 1B Workbook*
Guide: *Primary Mathematics 1B, Standards Edition, Home Instructor's Guide* (this book)
Extra Practice: *Primary Mathematics, Standards Edition, 1*
Tests: *Primary Mathematics, Standards Edition, 1B Tests*

Week		Objectives	Text book	Work book	Guide
Unit 11: Comparing Numbers					
		Chapter 1: Comparing numbers			1
1	1	♦ Compare two numbers. ♦ Find the number that is 1 more or 1 less than a given number within 10. ♦ Use *more than* and *less than*.	8-11	7-10	2-3
		Extra Practice, Unit 11, Exercise 1A, pp. 99-100			
		Tests, Unit 11, 1A and 1B, pp. 1-6			
		Chapter 2: Comparison by Subtraction			4-5
	1	♦ Compare two sets of objects to find how many more or less one set is than the other.	13-15	11-12	6
		Extra Practice, Unit 11, Exercise 1B, pp. 101-102			
	2	♦ Compare two sets with subtraction.	12-15	13-16	7-8
		Extra Practice, Unit 11, Exercise 2, pp. 103-104			
	3	♦ Compare two numbers with subtraction. ♦ Review addition and subtraction facts within 10.		17-18	9
		Tests, Unit 11, 2A and 12B, pp. 7-10			
		Tests, Unit 11 Cumulative Tests A and B, pp. 11-17			
Unit 12: Graphs					
		Chapter 1: Picture Graphs			10
2	1	♦ Read and interpret data in a simple picture graph.	16-18	19-23	11-12
	2	♦ Understand tally charts and bar graphs. ♦ Interpret simple bar graphs.	19-21	24-29	13-14
		Extra Practice, Unit 12, Exercise 1A and 1B, pp. 109-114			
	3	♦ Gather data and construct simple bar graphs.			15
		Tests, Unit 12, 1A and 1B, pp. 19-24			
		Tests, Unit 12 Cumulative Tests A and B, pp. 25-31			

Week		Objectives	Text book	Work book	Guide
Unit 13: Numbers to 40					
		Chapter 1: Counting			16
	1	♦ Count within 40 by making groups of ten first. ♦ Read numerals to 40.	22-26	30-31	17-18
	2	♦ Read and write number words to 40.		32-33	19
		Extra Practice, Unit 13, Exercise 1A, pp. 123-124			
3	3	♦ Express 2-digit numbers as tens plus ones.	26-27	34-36	20
	4	♦ Count on and count back 1, 2, or 3 within 40.	28	37-38	21
	5	♦ Order numbers within 40.	29	39	22-23
		Extra Practice, Unit 13, Exercise 1B, pp. 125-126			
		Tests, Unit 13, 1A and 1B, pp. 33-35			
		Chapter 2: Tens and Ones			24
	1	♦ Interpret a 2-digit number in terms of tens and ones.	30-33	40-41	25
	2	♦ Count on or back 1 or 10.	34-35	42-44	26
		Extra Practice, Unit 13, Exercise 2, pp. 127-128			
		Tests, Unit 13, 2A and 2B, pp. 37-39			
		Chapter 3: Addition and Subtraction			27-28
4	1	♦ Add a 1-digit number to a 2-digit number within 40 (no renaming). ♦ Subtract a 1-digit number from a 2-digit number within 40 when there are enough ones (no renaming).	36-37	45-46	29-30
	2	♦ Add or subtract a 1-digit number.	38	47-48	31
	3	♦ Add 1, 2, or 3 to a number within 40 by counting on. ♦ Subtract 1, 2, or 3 from a number within 40 by counting back.	39	49-50	32-33
	4	♦ Review: Add within 20 by making a ten.			34
	5	♦ Add a 1-digit number to a 2-digit number within 40 with renaming by making tens.	39-41	51-54	35-36
5	6	♦ Add a 1-digit number to a 2-digit number within 40 by using addition facts and renaming.		55-57	37-38
		Extra Practice, Unit 13, Exercise 3A, pp. 129-132			

Week		Objectives	Text book	Work book	Guide
	7	♦ Review: Subtract within 20 by subtracting from a ten.		58	39
	8	♦ Subtract ones from tens.	42-43	59-60	40
	9	♦ Subtract a 1-digit number from a 2-digit number within 40 with renaming.	42-43	61-62	41-42
		Extra Practice, Unit 13, Exercise 3B, pp. 133-136			
		Tests, Unit 13, 3A and 3B, pp. 41-43			
		Chapter 4: Adding Three Numbers			43
6	1	♦ Add three 1-digit numbers.	44-45	63-65	44-45
		Extra Practice, Unit 13, Exercise 4, pp. 137-138			
		Tests, Unit 13, 4A and 4B, pp. 45-47			
		Chapter 5: Counting by 2's			46
	1	♦ Count forward by twos. ♦ Count backward by twos.	46-47	66-67	47-48
		Extra Practice, Unit 13, Exercise 5, pp. 139-140			
		Tests, Unit 13, 5A and 5B, pp. 49-52			
		Review		68-73	49
		Tests, Unit 13 Cumulative Tests A and B, pp. 53-58			
Unit 14: Multiplication					
		Chapter 1: Adding Equal Groups			50
7	1	♦ Understand the concept of equal groups. ♦ Find the total number of objects in a given number of equal groups.	48, 50	76-77	51
	2	♦ Use mathematical language such as "3 fives" or "3 groups of 5" to describe equal groups. ♦ Find the total number in equal groups by repeated addition.	49, 51	74-75 78-79	52
		Extra Practice, Unit 14, Exercise 1, pp. 143-144			
		Tests, Unit 14, 1A and 1B, pp. 59-62			
		Chapter 2: Making Multiplication Stories			53
	1	♦ Write multiplication equations using "x" and "=" for a given situation involving multiplication.	52-55	80-81	54
	2	♦ Interpret multiplication equations with manipulatives or illustrations.		82	55
		Extra Practice, Unit 14, Exercise 2, pp. 145-146			
		Tests, Unit 14, 2A and 2B, pp. 63-69			

Week		Objectives	Text book	Work book	Guide
		Chapter 3: Multiplication Within 40			56
8	1	♦ Multiply within 40 using repeated addition.	56-58	83-85	57
	2	♦ Use rectangular arrays to illustrate multiplication. ♦ Solve picture problems involving multiplication.	59	86-88	58
		Extra Practice, Unit 14, Exercise 3, pp. 147-148			
		Tests, Unit 14, 3A and 3B, pp. 71-77			
		Review		89-100	59
		Tests, Unit 14 Cumulative Tests A and B, pp. 79-86			
Unit 15: Division					
		Chapter 1: Sharing and Grouping			60
9	1	♦ Understand division as sharing. ♦ Use manipulatives to illustrate sharing.	60-61 63-64	101-104	61
	2	♦ Understand division as grouping. ♦ Use manipulatives to illustrate grouping.	60, 62 64-65	105-108	62
		Extra Practice, Unit 15, Exercise 1, pp. 151-152			
		Tests, Unit 14, 1A and 1B, pp. 87-90			
		Tests, Unit 15 Cumulative Tests A and B, pp. 91-98			
Unit 16: Making Halves and Fourths					
		Chapter 1: Making Halves and Fourths			63
	1	♦ Recognize halves and fourths.	66-67	109-112	64-65
		Extra Practice, Unit 16, Exercise 1, pp. 155-156			
	2	♦ Describe and continue a pattern of halves or fourths according to position.		113-114	66
		Tests, Unit 16, 1A and 1B, pp. 99-102			
		Tests, Unit 16 Cumulative Tests A and B, pp. 103-109			
Unit 17: Time					
		Chapter 1: Telling Time			67
10	1	♦ Tell time to the hour. ♦ Read and write times for the hour (e.g. "6 o'clock"). ♦ Relate time to the hour to events of the day.	68-69	115-117	68-69
	2	♦ Tell time to the half hour on an analog clock face. ♦ Write times for the half hour (e.g. "half past 4"). ♦ Relate time to the half-hour to events of the day.	70-71	118-120	70
	3	♦ Relate time to another time or to an even using before or after.	72	121-122	71
		Extra Practice, Unit 17, Exercise 1A and 1B, pp. 159-162			
		Tests, Unit 17, 1A and 1B, pp. 111-114			

Week			Objectives	Text book	Work book	Guide
		Chapter 2: Estimating Time				72
	1		♦ Estimate time to the nearest half-hour. ♦ Compare the time it takes to do specific activities.	73-75	123-124	73
			Extra Practice, Unit 17, Exercise 2, pp. 183-184			
			Tests, Unit 17, 2A and 2B, pp. 115-118			
11		**Review**			125-129	74
			Tests, Unit 17 Cumulative Tests A and B, pp. 119-126			
Unit 18: Numbers to 100						
		Chapter 1: Tens and Ones				75
	1		♦ Count by tens. ♦ Recognize number words for tens.	76-79	130-132	76
	2		♦ Count within 100 by making tens.	80-81	133-136	77
	3		♦ Read and write number words within 100.		137-138	78
	4		♦ Interpret 2-digit numbers as tens and ones in terms of a part-whole model.	82	139-140	79
			Extra Practice, Unit 18, Exercise 1, pp. 175-178			
			Tests, Unit 18, 1A and 1B, pp. 127-130			
		Chapter 2: Estimation				80
12	1		♦ Estimate quantities by comparing to a known quantity. ♦ Estimate quantity by making a reasonable guess.	83-84	141	81
			Extra Practice, Unit 18, Exercise 2, pp. 179-180			
			Tests, Unit 18, 2A and 2B, pp. 131-135			
		Chapter 3: Order of Numbers				82
	1		♦ Understand order of numbers. ♦ Compare numbers within 100 using a number chart.		142-143	83
	2		♦ Find the number that is 1 or 10 less or the number that is 1 or 10 more than a given number within 100.		144-146	84
	3		♦ Count on and count back by tens and ones.	85-88	147-149	85-86
			Extra Practice, Unit 18, Exercise 3, pp. 181-182			
			Tests, Unit 18, 3A and 3B, pp. 137-139			

Week		Objectives	Text book	Work book	Guide
		Chapter 4: Comparing Numbers			87
13	1	♦ Compare numbers within 100 by comparing tens and then ones. ♦ Use the symbols for greater than and less than.	89-90	150-152	88-89
		Extra Practice, Unit 18, Exercise 4, pp. 183-184			
		Tests, Unit 18, 4A and 4B, pp. 141-144			
		Chapter 5: Addition Within 100			90-91
	1	♦ Add a 1-digit number to a 2-digit number within 100 (no renaming).	91-92	153-154	92
	2	♦ Add a 1-digit number to a 2-digit number with renaming.	92	155-156	93-94
	3	♦ Add tens to a 2-digit number.	93-94	157-160	95
	4	♦ Add 2-digit numbers.	95	161-162	96
14	5	♦ Understand the vertical representation for addition problems.	96-99	163-166	97
		Extra Practice, Unit 18, Exercise 5, pp. 185-188			
		Tests, Unit 18, 5A and 5B, pp. 145-148			
		Chapter 6: Subtraction Within 100			98-99
	1	♦ Subtract a 1-digit number to a 2-digit number within 100 (no renaming).	100-101	167-168	100
	2	♦ Subtract a 1-digit number from a 2-digit number with renaming.	101-102	169-170	101-102
	3	♦ Subtract tens from a 2-digit number.	103-104	171-174	103
15	4	♦ Subtract 2-digit numbers.	105	175-176	104
	5	♦ Understand the vertical representation for subtraction problems.	106-109	177-180	105
		Extra Practice, Unit 18, Exercise 6, pp. 189-192			
		Tests, Unit 18, 6A and 6B, pp. 149-152			
		Review		181-185	106
		Tests, Unit 18 Cumulative Tests A and B, pp. 153-162			

Week			Text book	Work book	Guide
Unit 19: Money					
	Chapter 1: Bills and Coins				107
16	1	◆ Recognize and name coins and bills up to $20. ◆ Change a coin or bill for an equivalent set of coins or bills of a smaller denomination. ◆ Count by fives.	110-114	186	108
	2	◆ Count the amount of money in a set of coins. ◆ Understand the symbol ¢.	115, 116	187-188	109
	3	◆ Count the amount of money in a set of bills. ◆ Understand the symbol $. ◆ Make up a set of coins or bills for a given amount.	115, 117	189-190	110
	4	◆ Compare the amount of money in two or three sets of coins or bills. ◆ Compare the price of two items in cents or dollars.	118	191-192	111
		Extra Practice, Unit 19, Exercise 1, pp. 197-200			
		Tests, Unit 19, 1A and 1B, pp. 163-170			
	Chapter 2: Shopping				112
17	1	◆ Add or subtract amounts of money in cents or dollars.	119-121	193-195	113
		Extra Practice, Unit 19, Exercise 2, pp. 201-202			
		Tests, Unit 19, 2A and 2B, pp. 171-176			
	Review			196-208	114
	Tests, Unit 19 Cumulative Tests A and B, pp. 177-185				
Answers to Mental Math					116
Appendix		Mental Math			a1-a9
		Graph for unit 12 lesson 2			a10
		1-40 chart			a11
		1-100 chart			a12
		Letters for unit 16 lesson 1			a13
		Number cards			a14-a17
		Dot cards			a18-a20
		Square graph paper			a21

Manipulatives

It is important to introduce the concepts concretely, but it is not important exactly what manipulative is used. A few possible manipulatives are suggested here. The linking cubes and playing cards will be used at many levels of the *Primary Mathematics*.

Whiteboard and Dry-Erase Markers
A whiteboard that can be held is useful in doing lessons while sitting at the table (or on the couch). Students can work problems given during the lessons on their own personal boards.

Multilink cubes
These are cubes that can be linked together on all 6 sides. Ten of them can be connected to form tens so that you have tens and ones. You can use Legos™ or anything else that can be grouped into tens, but the multilink cubes will be useful when the students get to volume problems in *Primary Mathematics* 4.

Base-10 set
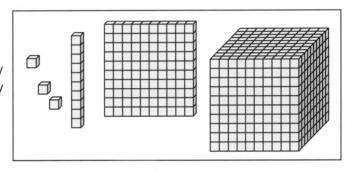
A set usually has 100 unit-cubes, 10 or more ten-rods, 10 hundred-flats, and 1 thousand-block. Since numbers in Primary Math 1 do not exceed 100, you can simply use multilink cubes instead, making ten-rods with them. For some activities, your student should not take the ten rods apart, but rather trade them in for unit cubes as appropriate.

Counters
Round counters are easy to use and pick up, but any type of counter will work.

Numbers sets
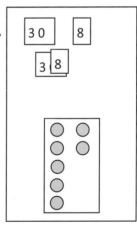
- Set 1: Numerals 0-9 and tens (10, 20, 30 ...). Use half an index card for 1-9, and whole index cards for 10-90 so that, when teaching numbers greater than 10, you can slide the single digits over the 10 card.
- Set 2: Dot cards for 0-10. You can use index cards and trace around round counters or pennies, or use dot stickers that are available as office supply. Have cards with 1, 2, 3, 4, or 5 down one side for the numbers 1-5, cards with 5 down one side and 1, 2, 3, 4, or 5 down the other side for numbers 6-10. Make 9 cards ones for the ten (9 cards with 10 dots on them).

Playing cards 0-10
You need four sets of number cards 0-10. You can use a deck of cards and call the ace 1. You can white out the ace and replace it with a 1. Use the cards to play games that help your student memorize the math facts. You can white out the J on the Jack and call that a 0, or use the tens from another deck and white out the 1 of the 10 and all the symbols.

Fact cards

A set of addition and subtraction fact cards (for addition and subtraction within 20). These can have the fact on one side and the answer on the other for individual practice.

Fact game cards: Fact cards with answers on separate cards rather than on the back of the card with the addition fact.

Hundred-chart

A number chart from 1-100. Laminated ones are nice, since you can use a dry-erase marker, or you can copy the one in the appendix.

Number cubes or dice

Some of the games or activities use number cubes. You can get blank cubes with labels, or simply label regular dice using masking tape. Several 10-sided dice are handy for games and activities, but not required.

Number chart 1-40 – Similar to hundred chart but for numbers 1-40. Use first four rows of a hundred-chart, or copy the chart in the appendix.

Clock – A real analog clock or a demonstration clock with geared hands.

Supplements

The textbook and workbook provide the kernel or essence of the math curriculum. Some students profit by additional practice and more review. Other students profit by more challenging problems. There are several supplementary books that can be used for either more practice, more challenge, or both. More information about these can be found at singaporemath.com. If you feel it is important that your student have a lot of drill in math facts, there are many websites that generate worksheets according to your specifications, or provide on-line fact practice. Web sites come and go, but doing a search using the terms "math fact practice" will turn up many sites. Playing simple games with cards is another way to practice math facts.

Unit 11 – Comparing Numbers

Chapter 1 – Comparing Numbers

Objectives

♦ Compare two numbers.
♦ Find the number that is 1 more or 1 less than a given number within 10.
♦ Use *more than* and *less than*.

Material

♦ Objects to count
♦ Numeral cards 1-10
♦ Linking cubes
♦ Counters
♦ Playing cards

Notes

In *Primary Mathematics* 1A students compared sets of objects concretely and pictorially where they counted the objects in each set and matched the objects in one set to the objects in the other set to see which set has more. This is reviewed here. In the next section, students will use subtraction to find out how many more objects are in one set than in the other. In Unit 18, they will learn the symbols '>' and '<'.

In *Primary Mathematics 1A* students also learned that whole numbers have a natural sequence which is known as order. The order is build on the concept of "1 more". This is reviewed here, using objects to illustrate "1 more" as "adding on" and "1 less" as "taking away."

Your student should be able to easily tell you what is one more or one less than any number through 20. In this unit and the next the numbers used are 10 or less. In the third unit, the concept of tens and ones is reviewed before going on to numbers greater than ten.

(1) Review: compare numbers

Textbook

Page 8

Tasks 1-9, pp. 9-11

1. No
2. Yes
3. No
4. (a) 4   (b) 4
5. 9
6. (a) 6   (b) 7
7. (a) 6   (b) 5
8. 9
9. (a) 8   (b) 8

Workbook

Exercise 1, pp. 7-8

1. (a) no
   (b) yes
   (c) yes
   (d) no
2. Ryan
3. Peter

Exercise 2, pp. 9-10

1. (a) 8   (b) 6
2. (a) 7   (b) 8
3. 1 less than 7 → 6
   1 more than 2 → 3
   1 less than 5 → 4
   1 less than 10 → 9
   1 more than 9 → 10
   1 more than 6 → 7
   1 more than 1 → 2

Activity

Use two types of objects, such as linking cubes and counters, or two types of toys.

Display two sets with different amounts. They do not need to be initially lined up — your student can do that while comparing them.

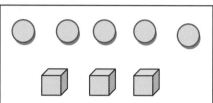

Ask your student to compare the sets. He might compare type (counter, block). Lead him to compare the number of objects, and to tell you which set has more objects. Then write the numbers.

Put the objects away and show him two number cards with the numbers for the previously shown sets. Ask your student to say which number is more, e.g., "5 is more than 3." Then, ask her to say which number is less, e.g., "3 is less than 5".

Show a set of objects, such as 7 linking cubes. Ask your student to count them. Ask her how many there will be if we **add** one more. Lead her to say, "8 is 1 more than 7". Repeat with a different set of objects.

Do a similar process for 1 less. Show your student a set of objects, such as 9 linking cubes. Ask him to count them, and then ask how many there will be if we take away, or **subtract** one. Lead him to say, "8 is 1 less than 9". Repeat with a different set of objects.

Show your student a number card. Ask her to write the number that is one more. Repeat with another card, this time asking her to write down the number that is one less. Repeat with other examples.

Discussion

Page 8

Tasks 1-3, p. 9

Write the numbers for each set in these tasks, and ask your student to compare the numbers.

Tasks 4-9, pp. 10-11

Encourage your student to write the answers.

Workbook

Exercises 1-2, pp. 7-10

Reinforcement

Game
<u>Material</u>: Deck of cards with face cards removed, or 4 sets o f number cards 1-10.
<u>Procedure</u>: Deal all cards out. Players turn over one card at a time and compare. The player with the highest number gets the card. If two players have the same number they turn over two more cards and compare them to see who wins the cards. The player with the most cards at the end wins.

Extra Practice, Unit 11, Exercise 1A, pp. 99-100

Tests

Tests, Unit 11, 1A and 1B, pp. 1-6

Chapter 2 – Comparison by Subtraction

Objectives

♦ Compare two sets of objects to find how many more or less one set is than the other.
♦ Compare two sets with subtraction.
♦ Compare two numbers with subtraction.
♦ Review addition and subtraction facts within 10.

Material

♦ Objects to count
♦ Numeral cards 1-10
♦ Linking cubes
♦ Counters
♦ Playing cards or several sets of numeral cards 1-10

Notes

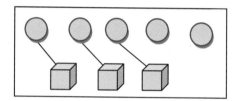

In *Primary Mathematics* 1A, Unit 4, students learned two subtraction situations: taking away and part-whole (missing part). In this unit students will learn another subtraction situation: comparison.

In comparison situations we compare two sets of objects to find out how many more or how many fewer objects are in one set than another. This can be done by matching one-to-one, or by subtraction.

When matching one-to-one students count the unmatched objects to find out how many more or how many fewer objects there are in one set than another.

Students will learn that to compare two sets, they can count the number of objects in each set, and then find the answer by subtraction: 5 – 3 = 2.

In this unit students will be practicing the concepts with numbers within 10. Comparison by subtraction will be extended to greater numbers in later units.

As you do this chapter continue to provide your student with opportunities to practice and use addition and subtraction facts within 10, if needed.

Your student may think in terms of addition to work out the subtraction problems; that is, find how much more must be added to the smaller number to make up the larger number. 3 and 2 make 5, so 5 – 3 = 2.

Students should be able to find the answers to the math facts shown on the next page with accuracy and reasonable speed by unit 3. Use flash cards, games, or drill sheets (which can be generated from numerous web sites). You can use the Mental Math 1-7 at the end of the guide. To use these, either point to the problem and have your student answer, or make copies and have your student write the answers. You could have your student do one mental math exercise a day for a while. They can be repeated.

0 + 0	0 + 1	0 + 2	0 + 3	0 + 4	0 + 5	0 + 6	0 + 7	0 + 8	0 + 9	0 + 10
1 + 0	1 + 1	1 + 2	1 + 3	1 + 4	1 + 5	1 + 6	1 + 7	1 + 8	1 + 9	
2 + 0	2 + 1	2 + 2	2 + 3	2 + 4	2 + 5	2 + 6	2 + 7	2 + 8		
3 + 0	3 + 1	3 + 2	3 + 3	3 + 4	3 + 5	3 + 6	3 + 7			
4 + 0	4 + 1	4 + 2	4 + 3	4 + 4	4 + 5	4 + 6				
5 + 0	5 + 1	5 + 2	5 + 3	5 + 4	5 + 5					
6 + 0	6 + 1	6 + 2	6 + 3	6 + 4						
7 + 0	7 + 1	7 + 2	7 + 3							
8 + 0	8 + 1	8 + 2								
9 + 0	9 + 1									
10 + 0										

0 − 0	1 − 1	2 − 2	3 − 3	4 − 4	5 − 5	6 − 6	7 − 7	8 − 8	9 − 9	10 − 10
1 − 0	2 − 1	3 − 2	4 − 3	5 − 4	6 − 5	7 − 6	8 − 7	9 − 8	10 - 9	
2 − 0	3 − 1	4 − 2	5 − 3	6 − 4	7 − 5	8 − 6	9 − 7	10 − 8		
3 − 0	4 − 1	5 − 2	6 − 3	7 − 4	8 − 5	9 − 6	10 − 7			
4 − 0	5 − 1	6 − 2	7 − 3	8 − 4	9 − 5	10 − 6				
5 − 0	6 − 1	7 − 2	8 − 3	9 − 4	10 − 5					
6 − 0	7 − 1	8 − 2	9 − 3	10 − 4						
7 − 0	8 − 1	9 − 2	10 − 3							
8 − 0	9 − 1	10 − 2								
9 − 0	10 − 1									
10 − 0										

(1) Compare two sets of objects

Textbook

Task 1, pp. 13-15

1. (a) 2   (b) 2
   (c) 2   (d) 5
   (e) 4   (f) 5

Activity

Show your student two sets of objects. The two sets should be different amounts. Get her to tell you which set has more.

Ask your student to line the objects up in pairs.

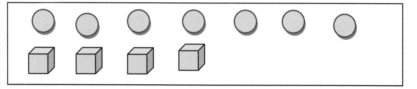

Then ask questions such as:

⇒ How many more counters are there than blocks?
⇒ How many fewer blocks are there than counters?

Repeat with other examples.

Draw a picture where two groups of the same item are separated, such as a pond with 8 frogs in it and 3 frogs on the bank. Ask your student whether there are more frogs on the bank or in the pond. He can draw lines to pair up a frog on the bank with a frog in the pond. Then ask him how many more frogs there are in the pond, and how many fewer frogs there are on the bank.

Discussion

Task 1, pp. 13-15

Have your student look at the pictures and answer the questions. Do not have her write a number sentence yet. She should be able to pair the objects visually, but if not, place counters on each picture, one color for one set and a different color for the other set of objects. Then let her pair the counters. She may wish to make up a story about the pictures.

Reinforcement

Let your student compare larger sets of objects, up to 20. You can use counters, multilink cubes, or toys. If you use two colors of multilink cubes, your student can link the pairs.

Extra Practice, Unit 11, Exercise 1B, pp. 101-102

Workbook

Exercise 3, pp. 11-12

Workbook

Exercise 3, pp. 11-12

1. (a) 3   (b) 2
   (c) 2   (d) 3
2. (a) Peter; 4
   (b) Lily; 4
   (c) 2   (d) 3

(2) Compare two sets with subtraction

Activity

Again, show your student two sets of objects and ask him to pair them up and tell you which set has more and how many more.

Tell your student that he can think of pairing the sets as each object in one set taking away an object in the other set. In the example illustrated here, each block "takes away" a counter. Move the pairs aside so that there are now two sets of counters – counters paired with planes and counters left over. Ask him what type of number equation we write to show that we take away some. We write a subtraction equation. Write the subtraction equation for this situation.

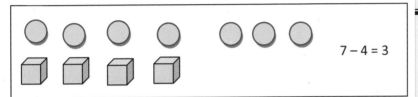

$$7 - 4 = 3$$

Point to each number in the equation and relate it to the situation. The first number is the number of objects in the larger set. The second is how many objects are in the smaller set. The answer is the **difference** between the two sets, or how many more cars there are than planes.

If your student has done *Primary Mathematics 1A*, she learned that subtraction can be thought of as taking away, or as finding a missing number for a number bond. You may want to illustrate some of the situations with a number bond. The total is the amount in the larger set. One part is the amount in the smaller set, each of which is paired with an item in the larger set. The missing part is the amount left over, that is, not paired. We use subtraction to find the missing part.

Repeat with another two sets of objects, but this time have your student write the subtraction equation.

Discussion

Page 11 | 2

Task 1, pp. 13-15

Ask your student to write a subtraction equation for each of the situations in task 1.

Textbook

Page 12

Task 1, pp. 13-15

1. (a) $5 - 3 = 2$
   (b) $5 - 3 = 2$
   (c) $6 - 4 = 2$
   (d) $9 - 4 = 5$
   (e) $7 - 3 = 4$
   (f) $8 - 3 = 5$

Workbook

Exercise 4, pp. 13-14

1. 2; 2
2. 4; 4
3. 2; 2
4. 4; 4

Exercise 5, pp. 15-16

1. 5; 5
2. 10; 10
3. $7 - 2 = 5$; 5
4. $10 - 7 = 3$; 3

Reinforcement

Let your student compare larger sets of objects, up to 20. Ask him to write a subtraction equation for each situation.

Extra Practice, Unit 11, Exercise 2, pp. 103-104

(3) Compare two numbers with subtraction

Activity

Show your students two numeral cards between 1 and 10, or write the two numbers down. They should be different numbers. Ask your student which is larger. Then ask him to write a subtraction equation to find how much larger. If necessary, you can illustrate the numbers with counters to begin with, but eventually you want your student to be able to determine which one is larger, how much larger, and write the subtraction equation without needing objects.

Repeat with another two numbers, this time asking your student which is smaller. Then ask her to write a subtraction equation to find how much smaller.

Ask your student if the answer also tells you how much larger one number is than the other. Point out that we can compare the two numbers by subtracting the smaller one from the larger one. The answer is how much larger one is than the other, or how much smaller.

Workbook

Exercise 6, pp. 17-18

Reinforcement

Use a deck of cards with face cards removed, or four sets of number cards 1-10. Shuffle the cards and lay down two at a time. Ask your student to tell you which is larger or smaller and by how much. If your student needs more practice writing subtraction equations, get him to write them for some of the pairs of numbers.

Extra Practice, Unit 11, Exercise 2, pp. 103-104

Enrichment

Have your student compare numbers within 20.

Tests

Tests, Unit 11, 2A and 12B, pp. 7-10

Tests, Unit 11 Cumulative Tests A and B, pp. 11-17

Workbook

Exercise 6, pp. 17-18

1. (a) $8 + 6 = 14$; 14
   (b) $8 - 6 = 2$; 2
2. $4 + 3 = 7$; 7
3. $8 - 2 = 6$; 6

Unit 12 – Graphs

Chapter 1 – Picture Graphs

Objectives

♦ Read and interpret data in a simple picture graph.
♦ Understand tally charts and bar graphs.
♦ Interpret simple bar graphs.
♦ Gather data and construct simple bar graphs.

Material

♦ Counters
♦ Small red, green, and purple counters or paper discs, about 1 cm in diameter (about ½ inch) or centimeter cubes
♦ Multilink cubes
♦ Stickers
♦ Appendix p. a10

Notes

A graph is a pictorial representation of data. It offers a visual display of relationships, making them more noticeable than if the data were presented simply in numerical form.

In *Primary Mathematics* 1B students will only deal with picture graphs and bar graphs based on a one-to-one correspondence, where one picture or one square corresponds to one item. In later levels they will deal with picture graphs with a scale, where picture in the graph represents more than one item and the bar graph has a numerical scale.

Students will first keep count of items by using counters, and then by using a tally chart. They will record the results in a picture graph or bar graph.

(1) Interpret picture graphs

Activity

Display four sets of objects, such as four sets of different color counters, and ask your student to compare them. Guide her in putting them in columns or rows, lining up the objects. Ask questions, such as

⇒ Which color do we have the most of?

⇒ How many more reds are there than greens?

⇒ How many fewer yellows are there than reds?

⇒ What color do we have the fewest of?

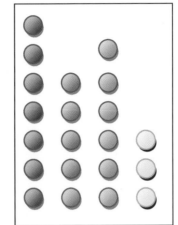

Point out that it is easier to compare the colors without having to count them if we put them in even rows or columns.

Tell your student that sometimes we can't line up the objects. For example, we might want to compare the number of different kinds of trees in a field. Instead, we can use counters to stand for the objects, and line up the counters instead.

Pages 16-17

Have your student do this activity. Get him to cover up each type of toy with the same color of 1 cm paper discs. Then move the discs over to the graph. Lead him to see that each color disc represents a toy and the number of toys of each kind can be compared by lining the discs up in a chart. Ask him to answer the questions on p. 17.

Discussion

Task 1, p. 18

Read the description at the top of the page and ask your student what sets we are comparing. We are comparing the number of books each child read last week. Tell her that the graph is called a **picture graph**. Ask your student to answer the questions on this page.

Tell your student that the graphs do not have to go up and down; they can go sideways. Look at the graphs in workbook on pages 21 and 22 for examples.

Textbook

Pages 16-17

1. 6
   2
   1
   14

Tasks 1, p. 18

1. (a) 3
   (b) Dave
   (c) 2   (d) 15

Workbook

Exercise 1, pp. 19-23

1. (a) 6   (b) 3
   (c) 5   (d) 14
   (e) 2
2. (a) 4   (b) 6
   (c) 3   (d) 2
3. (a) 8   (b) 6
   (c) 2   (d) 2
   (e) 18
4. (a) 20   (b) 5
   (c) 2   (d) 3
   (e) owls
   (f) eagles
5. (a) 7   (b) 4
   (c) 3   (d) 3
   (e) angelfish
   (f) goldfish

Workbook

Exercise 1, pp. 19-23

Reinforcement

Ask your student to write the addition or subtraction equation that would be used to answer the questions involving comparison. For example:

How many more cars than dolls are there? $5 - 3 = 2$

(2) Understand tally charts and bar graphs

Activity

Refer to the picture graph on p. 18 of the textbook. Ask your student for suggestions about how the children kept track of how many books they read. They could have kept a stack of the books, and then counted them at the end of the week. But perhaps some of the books had to go back to the library before the end of the week. Tell your student that one way to keep track of the books is to make a mark for each book, perhaps on a calendar, and then count the marks at the end of the week.

Draw a chart, and draw tally marks as you discuss tallying. For example, say, "Ali read a book on Monday," and make a mark for that. "Then he read a book on Wednesday," and make a mark for that. And so on. As you make the marks for Dave's books, tell your student that when we make the fifth mark,

Ali	///
Dave	////\ //
Rosni	////\

we draw it across the previous four. This makes it easier to count the marks since we know that they are in groups of 5. Count the marks for Dave by pointing at the set of 5 marks, saying "5", and counting on from that.

Knock on the table slowly and ask your student to keep track of how many times you are making a noise by drawing a tally mark for each knock. Continue for about 18 knocks, making sure she puts every 5th tally mark across the previous four. Then have her tell you how many times you knocked. Point out that although she could have simply counted, sometimes when we are collecting information to put into a table the things we are counting are not always one right after another, and it is easy to lose track, or forget the last number counted. Knock again and have her tally the knocks, but this time stop part way through and talk briefly about something else. Then ask her how many knocks you made so far. Then knock a few more times to continue the tallying.

Refer to problems 2 and 3 in workbook exercise 1, pp. 20-21. Ask your student how he answered 2(c). Point out that since the pictures of the animals are lined up, he could simply have counted the 3 more fish by counting up from the fish that lines up with the rabbit. Draw horizontal lines across the graph to make rows to show that the pictures are lined up. Now ask him how he answered 3(c). Point out that because the bear heads are larger than the monkey heads, the pictures don't line up, so it is not as

Textbook

Tasks 2-4, pp. 19-31

3. (a) Strawberries
    (b) 4
    (c) 3
    (d) 20

Workbook

Exercise 2, pp. 24-25

1. check graph
    Ali: 5
    Sally: 4
    Peter: 2
2. check graph
    Bears: 3   Cars: 3
    Bats: 6   Jump ropes: 7
    Drums: 4

Exercise 3, pp. 26-29

1. Circle: ////\
    Square: ////
    Triangle: ////\ /
    (a) 5   (b) 4
    (c) 6   (d) triangle
    Big: ////\ ////\
    Small: ////\
    (a) 10   (b) 5
    (c) 5
2. (a) 4   (b) dogs
    (c) bird
3. Dog: ////\ ///
    Bird: ///
    Rabbit: ////
    Cat: ////\ /
    (a) 3   (b) birds

easy to compare the two amounts without counting both. Tell him that normally, in a picture graphs, the pictures are lined up in rows across so that the amounts are easier to compare.

Tell your student that, rather than drawing picture graphs, we can draw a different kind of graph, called a bar graph. Use graph paper to create a bar graph for the data in problem 3 of the workbook on p. 21, or use the one in the appendix of this guide. Get your student to color in one square for each animal. Tell her that this type of graph is called a **bar graph**. Point out that each filled in rectangle in the graph stands for an item and the different items are lined up across so it is easy to compare the amounts for each.

Discussion

Tasks 2-4, pp. 19-21

Point out that when counting objects that are all over the place in a picture, we can put little check marks next to each picture as we count or write the tally marks so that we do not count the same things twice. You may also want to point out, when discussing the bar graph on p. 20, that in this case it is just as easy to color in one square for each fruit checked, rather than first creating a tally chart. But other times, when we collect data to put in charts, it is useful to do a tally chart. In task 4, children went around and asked others what their favorite ice-cream flavor was. In this case, making a tally chart is a good way to keep track of the answers.

Workbook

Exercises 2-3, pp. 24-29

Reinforcement

Extra Practice, Unit 12, Exercise 1A and 1B, pp. 109-114

(3) Create bar graphs

Activity

Note: Rather than creating bar graphs as a separate math activity, it might be better to simply integrate graphs into other curriculum areas, such as science. Students can learn about graphs and how to interpret them in math, and then create them and use them for a specific purpose, rather than for some made-up purpose. Most science curricula require students to create graphs. However, if you would like to give your student experience in creating a graph now, you can do the following activity.

Provide your student with 3 to 5 sets of objects with up to 20 objects in each set, a large sheet of paper divided into columns, and stickers or paper shapes that can be stuck onto the graph. Help your student create a picture graph or bar graph to compare the numbers in each set of objects.

Reinforcement

Have your student collect data of some sort where there are less than 20 for each set, such as the number of books read each month for 4 months, the number of times the phone rang for a particular member of the household in a day, the number of sunny versus cloudy versus rainy days for two weeks, or something else of interest. Have your student keep track of the data using tally marks. Then, help her create a picture graph for the data and discuss what the graphs shows.

Tests

Tests, Unit 12, 1A and 1B, pp. 19-24

Tests, Unit 12 Cumulative Tests A and B, pp. 25-31

Unit 13 – Numbers to 40

Chapter 1 – Counting

Objectives

- Count within 40 by making groups of ten first.
- Read numerals to 40.
- Read and write number words to 40.
- Express 2-digit numbers as tens plus ones.
- Count on and count back 1, 2, or 3 within 40.
- Order numbers within 40.

Material

- Multilink cubes
- Numeral cards
- Dot cards
- Number cards, 1-40
- Number chart, 1-40 (see appendix p. a11)

Notes

In *Primary Mathematics* 1A students learned to interpret numbers from 11 to 20 in terms of a ten and ones.

In this chapter your student will interpret numbers from 20 to 40 in terms of tens and ones. Although most students can count to high numbers by rote, they need to have a good grasp of place-value and the base-10 number system. In this and the next section you want to emphasize place-value and the meaning of the digits in each place, the tens place and the ones place. You can use the part-whole number bond model from *Primary Mathematics* 1A to show that a 2-digit number is made up of tens and ones.

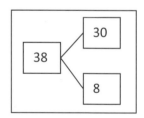

Give your student as much practice as needed counting and making tens with units, such as straws, toothpicks, craft sticks, or multilink cubes, before using coins or base-10 blocks.

Your student will also review order in this section and will compare 2-digit numbers by first comparing the tens, and, if they are the same, then comparing the ones.

Your student will be counting by making groups of 10, using objects that are easy to put into groups. Make sure he understands that he is doing this in order to understand written numbers and what the digits in different places represent. We cannot physically group everything that we count into tens and ones. Thinking in terms of tens and ones is for the purpose of *writing* amounts greater than 9. More than 9 ones are "allowed" in a column of place-value chart when adding or subtracting with renaming (in *Primary Mathematics* 2), but in order to show the number in the place value chart as it is written, we make groups of ten with the manipulatives (or trade in 10 ones for a ten) if we do have more than 9 ones.

(1) Count to 40 by making groups of ten

Activity

Give your student 18 multilink cubes and ask him to count them. (Start with 11 cubes instead if your student still needs help with counting the teen numbers.) After he counts them, remind him, if needed, that we can count by first making groups of ten. Have your student link 10 cubes in two rows of 5. Write the number 18. Remind your student that the numbers 0-9 are called **digits**. In order to write a number greater than 9, we have to make a group of ten and write the number with 2 digits, one for the groups of ten, and one for the ones. 18 is a 2-digit number. The first digit tells us how many tens there are, and the next digit tells us how many ones there are. You can show this with the numeral cards (sliding the 8 over the 10) and/or with a number bond.

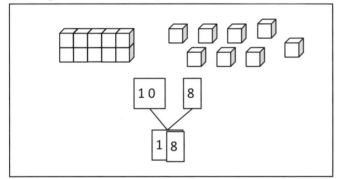

Add another cube and ask your student to count on as you add them: 19, 20. Have her make another linked set of 10. Write the number 20. Ask her for the meaning of each of the digits. The 2 tells us that there are 2 groups of 10, and the 0 tells us that there are no ones. Ask her why we need the 0, if there are no ones? It keeps the 2 in the right place; without it we would not know if the 2 is for the number of tens or the number of ones.

Guide your student in adding more blocks and counting to 30, and then to 40. Write the numbers as he says them in a column so he can see the pattern. The ones digit increases from 1 to 9, then the tens digit becomes 3, then the ones digit again increases from 1 to 9, and then the tens digit becomes 4. At each ten link the cubes into a ten.

Have your student read the list of numbers. You can extend it back to 1.

18
19
20
21
22
23
24
25
26
27
28
29
30
31
32
33
34
35
36
37
38
39
40

Textbook

Page 22-24

Tasks 1-5, pp. 24-26

1. 25
2. 28
3. 30
4. 34
5. 40

Workbook

Exercise 1, pp. 30-31

1.	24	27
	30	33
	36	37
2.	38	5; 15
	4; 24	2; 32
3.	20	26; 26
	38; 38	29; 29

Discussion

Pages 22-24

Tasks 1-5, pp. 24-26

These pages include the number words; you can save the number words for the next lesson. Guide your student in reading the numbers in the chart at the top of p. 25 in order. The numbers in the first row are read first, then in the second row, starting back at the left side, and so on.

Reinforcement

Use number cards 1-40. Mix them up and ask your student to put them in order.

Use a large sheet of paper. Draw 10 straight lines or get your student to draw ten straight lines to divide the paper into parts. Then get your student to count the parts and write the numbers in the spaces as she counts.

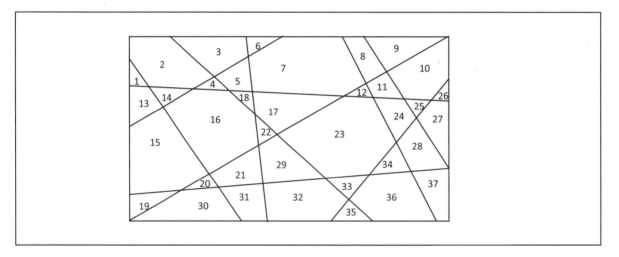

It can be helpful for students to be able to visualize pattern in the chart at the top of page 25. You can create a large chart showing the numbers along with their representation as tens and ones using drawings of the multilink cubes or dot cards. Your student might enjoy creating it and/or coloring it.

(2) Read number words

Activity

Make sure your student can read the number words:

zero	ten	twenty
one	eleven	thirty
two	twelve	forty
three	thirteen	
four	fourteen	
five	fifteen	
six	sixteen	
seven	seventeen	
eight	eighteen	
nine	nineteen	

Tell your student that for number words past twenty, we write a hyphen (-) between the word for the tens and the word for the ones.

Ask your student to read the number words in the textbook on pages 23-26.

Write some number words, and have your student write the numerals.

If you want your student to learn to spell the number words at this time, have him practice writing them, and then say the numbers and ask him to write the number words.

Reinforcement

Use number word cards for 1-40. Mix them up and ask your student to put them in order.

Optional: Have your student glue ten beans in a row on ten craft stick for another representation of tens and ones that can be used in later lessons.

Extra Practice, Unit 13, Exercise 1A, pp. 123-124

Textbook

Pages 23-26

Workbook

Exercise 2, pp. 32-33

1.	22		38
	31		37
	35		26
	23		40
	34		29
2.		36	
	25	39	32
	30	34	28
	27	24	33

(3) Express 2-digit numbers as tens plus ones

Textbook

Tasks 6-8, pp. 26-27

6. (a) 26
   (b) 26
   (c) 26
7. (a) 38
   (b) 38
   (c) 38
8. (a) 24
   (b) 27
   (c) 35

Workbook

Exercise 3, p. 34

1. (a) 21
   (b) 33
   (c) 27
   (d) 35

Exercise 4, pp. 35-36

1. (a) 4
   (b) 8
   (c) 30
   (d) 36
   (e) 39
2. 25   37
   6   3
   20   30

Activity

Show your student some tens and ones, and ask her questions such as:

⇒ 3 tens and 6 ones make what?
⇒ What is 6 more than 30?
⇒ 30 and what is 36?

Write a 2 digit number within 40 and ask your student to finish a number bond showing it as tens and ones, where the tens, the ones, or the whole is missing.

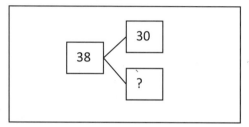

Discussion

Tasks 6-8, pp. 26-27

Reinforcement

use a variety of material that can be grouped into tens, such as the multilink cubes, dot cards, craft sticks or straws taped in bundles of tens, ten beans glued on a craft stick. It is important that your student does not associate tens and ones only with one type of manipulative – all objects are essentially grouped into tens and ones in order to represent the amount with digits (even if they can not physically grouped).

Write a number between 20 and 40 and have your student read the number and illustrate with the tens and ones using the multilink cubes or dot cards or other groups of tens and ones.

(4) Count on and count back 1, 2, or 3 within 40

Activity

Use a number chart for 1-40. Cover up some numbers and get your student to name and write the missing numbers.

1	2	3	4	5	6	7	8	9	10
11	12	13	14	15	16	17	18	19	20
21	22	23	24	25	26	27	28	29	30
31	32	33	34	35	36	37	38	39	40

Use the same number chart. Name a number and ask your student to answer questions such as the following:

⇒ What number is between 24 and 26?
⇒ What numbers are between 29 and 35?
⇒ What number is 1 less than 31?
⇒ What number is 3 more than 28?

In particular, ask questions that make your student go to the next or previous row.

Use number cards 1-40. Pick out a set of about 10 or so consecutive numbers. Remove several random numbers between the smallest and the largest. Mix up the remaining cards and ask your student to put them in order and find which numbers are missing.

Discussion

Tasks 9-10, p. 28

Workbook

Exercise 5, pp. 37-38

Reinforcement

Draw a 6 by 6 array, or use the blank squares on the back of a laminated hundreds board. Fill the squares in randomly with numbers between 1 and 40, leaving four random numbers out. Ask your student to determine which ones are missing.

10	18	13	2	28	36
27	26	1	19	3	29
17	9	5	6	25	12
21	14	20	30	7	24
31	33	4	22	32	40
37	15	16	8	11	38

Enrichment

Write down a number and ask your student what number is 1, 2, or 3 more or less than that number, without letting him use a number chart to find the answer.

Textbook

Tasks 9-10, p. 28

9. 23, 25, 26, 29, 32, 34, 35, 38, 40

10. (a) 25
    (b) 29
    (c) 38
    (d) 26

Workbook

Exercise 5, pp. 37-38

1. (a) 13, 14; 16; 18; 20
   (b) 22, 23; 25; 28; 29
   (c) 33, 34; 36; 37; 39, 40

2. (a) 16
   (b) 27
   (c) 31
   (d) 17
   (e) 32
   (f) 39
   (g) 19
   (h) 31
   (i) 26
   (j) 35

(5) Order numbers within 40

Textbook

Tasks 11-12, p. 29

11. (a) 27
    (b) 36
12. (a) 37
    (b) 14
    (c) 14, 24, 34, 37

Workbook

Exercise 6, p. 39

1. (a) 21
       40
       40
       21
   (b) 17
       39
       39
       17

Activity

Use a number chart 1-40. Point to two numbers in the same row. Ask your student which one is greater. The one farther to the right, or with the greater number in the ones digit, is greater. Repeat with two numbers in the same row. Point out that the one farther down, or with the greater number in the tens digit, is greater. Repeat with two numbers in different columns and rows, such as 28 and 32. The one with the larger tens is greater.

1	2	3	4	5	6	7	8	9	10
11	12	13	14	15	16	17	18	19	20
21	22	23	24	25	26	27	28	29	30
31	32	33	34	35	36	37	38	39	40

Get your student to show you the two numbers with tens and ones in multilink cubes or dot cards. The one with the most tens is greater, even if it has fewer ones.

Write down the numbers 38 and 34 and then write them one above the other. Ask which one is greater and which one is smaller. Point out that the tens are the same, but the ones are not, and the number with more ones, 38, is greater.

```
38, 34
3 8     3 tens   8 ones
3 4     3 tens   4 ones
```

Repeat with 24 and 44. This time, the number with the higher tens digit, 44, is the greater number.

```
24, 44
2 4     2 tens   4 ones
4 4     4 tens   4 ones
```

Repeat with 32 and 28. Make sure your student understands that if the tens are different, all we need to compare are the tens.

```
32, 28
3 2     3 tens   2 ones
2 8     2 tens   8 ones
```

Repeat with 9 and 24. Since the 1-digit number essentially has 0 tens, it is smaller. Make sure your student understands that we are not just comparing the first number, but the tens.

```
9, 24
  9     0 tens   9 ones
2 4     2 tens   4 ones
```

Discussion

Tasks 11-12, p. 29

For task 12, lead your student to see that she can first put in order by the tens, and then if two or more have the same tens, then order by the ones. Write the numbers on index cards so that she can move them around to arrange them in order.

Workbook

Exercise 6, p. 39

Reinforcement

Use number cards 1-40. Mix them up and pick one out randomly. Ask your student to guess the number. She can ask questions that can be answered by "yes" or "no" only, such as, "Is it greater than 20?"

Extra Practice, Unit 13, Exercise 1B, pp. 125-126

Tests

Tests, Unit 13, 1A and 1B, pp. 33-35

Chapter 2 – Tens and Ones

Objectives

♦ Interpret a 2-digit number in terms of tens and ones.
♦ Count on or back 1 or 10.

Material

♦ Base-10 blocks, tens and ones
♦ Multilink cubes
♦ Craft sticks, rubber bands
♦ Dimes and pennies [10-cent and 1-cent coins]

Notes

This chapter continues the study of place-value. A thorough understanding of place value is necessary and will help students learn the computational algorithms of the four operations – addition, subtraction, multiplication, and division – in a meaningful way.

Your student should be able to model 2-digit numbers as tens and ones in a variety of ways; number bonds, number cards where the ones overlap the tens, and on a place-value chart. The place-value chart will be useful as more places (e.g., hundreds, tenths) are added. At this level, the place-value chart has two columns, one labeled for tens and one for ones. When representing a 2-digit number in a place-value chart, the tens goes in the tens column, and the ones in the ones column. In *Primary Mathematics 2*, students will be placing number discs (e.g., discs labeled with "10" or "1") on the chart to aid them in understanding the computational algorithms.

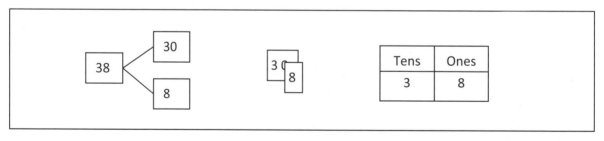

Your student will be adding 1 or 10 to a number within 40 and subtracting 1 or 10 from a number within 40. She need to recognize which place they are adding to or subtracting from.

(1) Interpret a 2-digit number in terms of tens and ones

Discussion

Pages 30-31

Ask him to tell you the number of tens and ones in each set of items (sticks, square tiles, money). Remind your student that 10 pennies [1-cent coins] is worth the same as a dime [10-cent coin] and we are finding the amount of money, not the amount of coins. Point out that all three represent the same number of sticks, square tiles, or cents.

Ask him what the digit 3 in the numeral 34 represents (3 tens) and what the digit 4 represents (4 ones). Then ask him how many ones in all there are in 34 (34 ones)

34 = 3 tens + 4 ones = 34 ones

Tasks 1-3, pp. 31-33

Workbook

Exercise 7, pp. 40-41

Reinforcement

Draw a large place-value chart for tens and ones.

Use base-10 blocks or multilink cubes or craft sticks in tens and ones. Write a number and have your student place the correct number of tens and ones in the correct columns of the chart.

Put tens and ones on the chart and then get your student to write the numeral.

Repeat with other base-10 representations, such as craft sticks with 10 beans glued on them and individual beans.

Textbook

Pages 30-31

Tasks 1-3, pp. 31-33

1. (a) 2 tens and 9 ones
   (b) 4 tens and 0 ones
2. (a) 30
   (b) 23
   (c) 38
3. (a) 24
   (b) 34
   (c) beads, 10

Workbook

Exercise 7, pp. 40-41

1. (a) 2 tens 3 ones
        2 tens 8 ones
        2 tens 9 ones
        2 tens 6 ones
        3 tens 8 ones
        3 tens 0 ones
2. (a) 2, 5 → 25
   (b) 3, 7 → 37
   (c) 4, 0 → 40

(2) Count on or back 1 or 10

Textbook

Tasks 4-5, pp. 34-35

4. (a) 25
   (b) 34
   (c) 23
   (d) 14
5. (a) 30
   (b) 39
   (c) 40
   (d) 20

Workbook

Exercise 8, pp. 42-44

1. (a) 19
   (b) 33
   (c) 22
   (d) 40
2. (a) 27
   (b) 26
   (c) 30
   (d) 21
3. 26 ; 35   21 ; 12
   25 ; 34   28 ; 37
   25 ; 16   28 ; 19

Activity

Use base-10 blocks or multilink cubes in tens and ones. Do an activity similar to the one pictured on p. 30 of the textbook: Use the base-10 material to display the tens and ones of the number 28. Ask your student to write the number. (It is important in this activity that your student now only say the number, but also write them.) Then get your student to show and say the number that is one more by adding a one to the ones. "One more than 28 is 29" and then writing the number. Then have him add another one. He can find the answer by counting up, "29, 30, One more than 29 is 30." Have him replace the ten ones with a ten and write the number down. Then ask him to show and write one more, 31.

Now ask her to show and write the number that is 10 more. Make sure she realizes that she can add a ten, not ten ones.

Now set out 2 tens and 2 ones. Ask your student to show and write the number that is one less, 21, and then again, 20, and then again, 19. To show 19, he will have to trade in a ten for 9 ones.

Now ask her to show and write the number that is ten less, 9, by taking away a ten.

Repeat with other numbers within 40 as needed.

Discussion

Tasks 4-5, pp. 34-35

Workbook

Exercise 8, pp. 42-44

Reinforcement

Write some addition expressions involving adding or subtracting 1 or 10, and ask your student to solve them.

$\Rightarrow$ 28 + 1
$\Rightarrow$ 30 – 1
$\Rightarrow$ 18 + 10
$\Rightarrow$ 22 – 10
$\Rightarrow$ 30 + 10

Extra Practice, Unit 13, Exercise 2, pp. 127-128

Tests

Tests, Unit 13, 2A and 2B, pp. 37-39

Chapter 3 – Addition and Subtraction

Objectives

- Add or subtract a ten from a 2-digit number.
- Add 1, 2, or 3 to a number within 40 by counting on.
- Subtract 1, 2, or 3 from a number within 40 by counting back.
- Add a 1-digit number to a 2-digit number.
- Subtract a 1-digit number from a 2-digit number within 40.

Material

- Base-10 blocks
- Multilink cubes
- Counters
- Hundred-chart
- Mental Math 1-17 (Appendix)

Notes

In *Primary Mathematics* 1A students learned the addition and subtraction facts within 20 through various strategies.

⇒ Add the ones: To add ones from a ten and ones, add ones to ones.

$$13 + 5 = 10 + 8 = 18$$
$$/\,\backslash$$
$$10\ \ 3$$

⇒ Make 10: If adding ones will give an answer greater than 10, split one of the numbers up in order to make a ten with the other number.

$$8 + 5 = 10 + 3 = 13$$
$$/\,\backslash$$
$$2\ \ 3$$

⇒ Subtract from the ones: To subtract ones from a ten and ones when there are enough ones, subtract the ones from the ones.

$$18 - 5 = 10 + 3 = 13$$
$$/\,\backslash$$
$$10\ \ 8 - 5 = 3$$

⇒ Subtract from the 10: To subtract ones from a ten and ones where there are not enough ones, subtract the ones from the tens and add that result to the ones.

$$13 - 8 = 3 + 2 = 5$$
$$/\,\backslash$$
$$3\ \ 10 - 8 = 2$$

⇒ Count on: When one of the numbers is 1, 2, or 3, count on from the other number to find the answer.

⇒ Count back: When the number being subtracted is 1, 2, or 3, count back from the other number to find the answer.

If your student has not learned these strategies, you should teach them as you proceed with this unit. Even if your student already knows all the addition and subtraction facts within 20, knowing the strategy of making the next higher place or subtracting from the next higher place will be useful for mental math strategies in a variety of situations, and increases your student's number sense and understanding of place value. Plus, if your student forgets a fact, or is unsure of one, these strategies offer a quick way to mentally compute the fact.

In this chapter the following strategies are used for addition and subtraction within 40:

⇒ Add, without renaming, using the strategy of adding the ones.

$$33 + 5 = 30 + 8 = 38$$
$$/\backslash$$
$$30 \quad 3$$

⇒ Subtract, without renaming, using the strategy of subtracting the ones.

$$38 - 5 = 30 + 3 = 33$$
$$/\backslash$$
$$30 \quad 8$$

⇒ Add, with renaming, by making a ten.

$$28 + 5 = 30 + 3 = 33$$
$$/\backslash$$
$$2 \quad 3$$

⇒ Subtract, with renaming, by subtracting from the ten.

$$33 - 8 = 3 + 22 = 25$$
$$/\backslash$$
$$3 \quad 30 - 8 = 22$$

⇒ Add when the 1-digit number is 1, 2, or 3 by counting on.

⇒ Subtract when the 1-digit number is 1, 2, or 3 by counting back.

(1) Add or subtract a 1-digit number, no renaming

Activity

Use base-10 material to illustrate the following addition situations, in order. Get your student to verbalize the process by using the words tens and ones. Write the addition equations. Keep them as a list of equations to discuss later.

3 + 4 = 7
3 ones + 4 ones = 7 ones

13 + 4 = 17
1 ten 3 ones + 4 ones = 1 ten 7 ones

23 + 4 = 27
2 tens 3 ones + 4 ones = 2 tens 7 ones

33 + 4 = 37
3 tens 3 ones + 4 ones = 3 tens 7 ones

Have your student look at the list of equations and ask if he sees any pattern. The tens stay the same in the answer, and only the ones change. We are adding ones to ones.

Textbook

Pages 36-37

5; 15; 25; 35
2; 12; 22; 32

Workbook

Exercise 9, pp. 45-46

1. (a) 6; 16; 26
   (b) 7; 27; 37
2. (a) 4; 14; 24
   (b) 0; 20; 30

Use base-10 material to illustrate the following subtraction situations, in order. Again, get your student to verbalize the process by using the words tens and ones. Write the addition equations and keep them as a list to discuss later.

8 – 3 = 5
8 ones – 3 ones = 5 ones

8 – 3 = 5

18 – 3 = 15
1 ten 8 ones – 3 ones = 1 ten 5 ones

18 – 3 = 15

28 – 3 = 25
2 tens 8 ones – 3 ones = 2 tens 5 ones

28 – 3 = 25

38 – 3 = 35
3 tens 8 ones – 3 ones = 3 tens 5 ones

Have your student look at the list of equations. Again, the tens stay the same in the answer, and only the ones change. We are subtracting ones form ones.

38 – 3 = 35

Discussion

Pages 36-37

Workbook

Exercise 9, pp. 45-46

Review

Mental Math 1-7

(2) Add or subtract a 1-digit number or ten

Activity

Use base-10 material or multilink cubes to illustrate a problem such as 27 + 3 where adding the ones makes a ten:

2 tens 7 ones + 3 ones = 2 tens 10 ones = 2 tens 1 ten

Ask your student how we would write the number for 2 tens 1 ten. Since we now have 3 tens, the answer is 30.

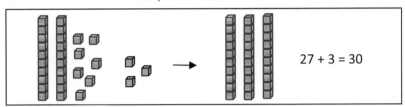

27 + 3 = 30

Represent the problems below with a number bond, discussion how we add 5 ones to ones and the ten to tens, and we subtract ones from ones and a ten from the tens. Use manipulatives if needed.

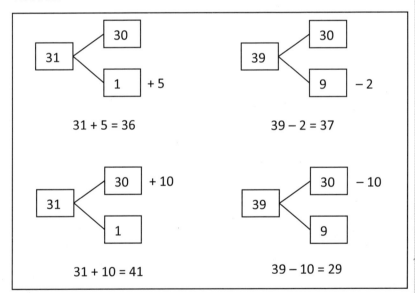

31 + 5 = 36   39 − 2 = 37

31 + 10 = 41   39 − 10 = 29

Discussion

Task 1, p. 38

Workbook

Exercise 10, pp. 47-48

Reinforcement

Mental Math 8

Textbook

Task 1, p. 38

1. (a) 28, 20
   (b) 30, 24
   (c) 39, 19

Workbook

Exercise 10, pp. 47-48

1. (a) 18
   (b) 40
   (c) 36
   (d) 22
   (e) 40
2. (a) 10
   (b) 15
   (c) 34
   (d) 30
   (e) 30

(3) Add or subtract 1, 2, or 3 by counting on or counting back

Textbook

Tasks 2-3, p. 39

2. 32

3. 29

Activity

Draw a circle and put 26 counters in it. Set 3 counters beside it. Write the expression: 26 + 3. Have your student count as you add the three counters to the circle: 27, 28, 29. She can check the answer by counting the total counters in the bowl (by making tens and ones). Tell her that when we add a small number, like 1, 2, or 3, it is easy to count up to add. Point out that when we count up, we do not include the number we are starting with.

Now there are 29 counters. Write the expression 29 + 2 and set two counters beside the circle. Have your student count as you add the counters to the circle. Write the answer. This time we have another ten.

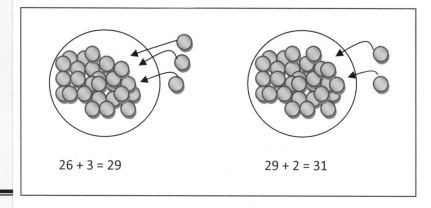

26 + 3 = 29   29 + 2 = 31

Use a similar process to illustrate 23 − 2 and then 21 − 3, this time counting back. Make sure your student does not include the number as one of the numbers counted back.

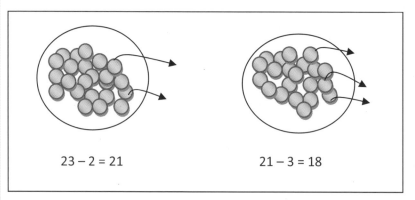

23 − 2 = 21   21 − 3 = 18

Workbook

Exercise 11, pp. 49-50

1. (a) 23   (b) 26
   (c) 30   (d) 32
   (e) 37   (f) 38
   (g) 34   (h) 40
2. (a) 22   (b) 25
   (c) 25   (d) 28
   (e) 31   (f) 36
   (g) 37   (h) 38
3. (a) 21   (b) 31
   (c) 19   (d) 29
   (e) 35   (f) 34
   (g) 33   (h) 30
   (i) 39   (j) 40
   (k) 36   (l) 37
4. 38; 36; 39; 40; 37

Use a number chart for 1-40. Point to a number and ask your student to add or subtract 1, 2, or 3 from it. He can put a counter on the number and count forward or back with it. Include problems where he has to go to the start of the next row or the end of the previous row to count on one or back one.

1	2	3	4	5	6	7	8	9	10
11	12	13	14	15	16	17	18	19	20
21	22	23	24	25	26	27	28	29	30
31	32	33	34	35	36	37	38	39	40

Discussion

Tasks 2-3, p. 39

Reinforcement

Label number cube with "+1", "+2", "+3", "−1", "−2", "−3". Write down the number 25. Have your student throw the cube, do the operation that comes up on the cube, and write the answer below the number 25. Repeat as many times as is helpful.

Mental Math 9

(4) Review: Add within 20 by making a ten

Activity

If necessary, review the "make 10" strategy for adding numbers within 20. Write a problem such as 7 + 6. Illustrate with base-10 material. Ask your student what is needed to make a ten with 7. Move the three ones over to the 7 ones and replace with a ten. Ask how many ones are left. Use number bonds to show that 6 has been split into 3 and 3. Write the answer first as 10 + 3 and then as 13.

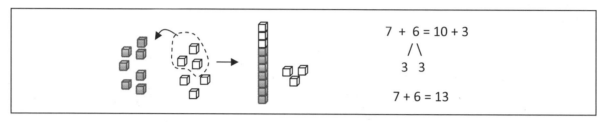

Provide additional examples.

Write some problems such as the following:

⇒ 8 + 4 = 10 + _____
⇒ 5 + 8 = _____

For the second problem, point out that we can either split up the 5 to add 2 to 8, or split up the 8 to add 5 to 5.

Write several problems showing the addition of 1-digit numbers, and ask your student to tell you if the answer will be greater than ten or less than ten, without finding the actual answer. For example:

⇒ 4 + 5 = _____
⇒ 6 + 3 = _____
⇒ 8 + 2 = _____
⇒ 3 + 2 = _____

Spend as much time on the idea of making a ten to add within 20 as needed.

Reinforcement

Mental Math 10-12

(5) Add a 1-digt number to a 2-digit number by making tens

Activity

Display 26 and 4 using base-10 material.

Write the problem: 26 + 4 and display 26 and 4 using base-10 material.

Ask your student to add the ones. 6 ones + 4 ones = 10 ones. Replace the ten ones with a ten-rod. Tell your student that since 6 and 4 make 10, by adding 4 to 26 we are making the next ten, or 30. Write the answer.

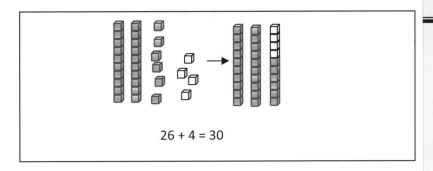

26 + 4 = 30

Now write the problem 26 + 8 and display 26 and 8. Ask your student how many we need to add to 26 to make the next ten, and move the 4 ones over. Trade in the 10 ones for a ten. Ask your student how many we now have. Since we took 4 of the ones to make a ten, there are 4 ones left, and we now have 3 tens and 4 ones. Add "= 30 + 4" to the addition expression. 26 and 8 is the same as 30 and 4. Then write: 26 + 8 = 34.

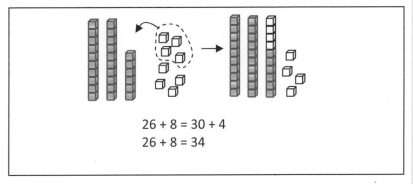

26 + 8 = 30 + 4
26 + 8 = 34

Textbook

Tasks 4-5, pp. 40-41

4. (a) 30
   (b) 33
5. (a) 31
   (b) 36
   (c) 40

Workbook

Exercise 12, pp. 51-52

1.	28	28
	38	29
	29	40
2.	28	35
	28	29
	37	38

Exercise 13, pp.43-54

1.	23	32
	33	31
	32	26
2.	21	34
	23	33
	32	40

Write the problem: 28 + ___ = 30.

After your student has found the answer, write the problem: 28 + 6 = 30 + ___. Have your student use the answer to the previous problem to solve this problem. Use base-10 material if needed. Repeat with other examples.

Write the problem: 22 + 3 = ____.

Ask your student if we would need to make the next ten with this problem. No, since the ones add to a number less than 10. Ask your student for the answer.

Write the problem: 22 + 9 = ____.

Ask your student if we need to make the next ten with this problem. Yes, since the ones add to a number greater than 10.

Discussion

Tasks 4-5, pp. 40-41

Workbook

Exercises 12-13, pp. 51-54

28 + _____ = 30
28 + 6 = 30 + _____

22 + 3 = _____
22 + 9 = _____

(6) Add a 1-digit number to a 2-digit number using addition facts

Activity

Have your students solve the following problems. Use base-10 material and number bonds to illustrate. Tell your student that we can first add the ten, then the ones.

$\Rightarrow$ 10 + 14
$\Rightarrow$ 20 + 14

Ask your student to find the answer to 8 + 6. Then, ask him to use that answer to find the answer to 28 + 6. Illustrate with base-10 material and number bonds.

8 + 6 = 14
28 + 6 = ?

28 + 6 = 20 + 14
/ \
20 8

28 + 6 = 34

Since 8 + 6 = 14, then
28 + 6 = 20 + 14 = 34.

Tell your student that if he knows 8 + 6, he can use this method rather than making the next ten (28 + 6 = 30 + 4). We add the ones together, but since the answer when adding the ones is more than ten, there will be another 10, so there is one more ten.

Illustrate some more examples. Tell your student he can use either method, making a ten or using the addition fact, to add.

Workbook

Exercises 14-15, pp. 55-57

Workbook

Exercise 14, p. 55

1. 15   16   16   15
   14   18   13   12
   17   11   15   14
   14   17   12   13

Exercise 15, pp. 56-57

1. 17
   7; 17   9; 19
   8; 28   9; 29
   9; 39   8; 38
2. 21
   10; 20   11; 21
   12; 32   12; 32
   14; 34   10; 40

Reinforcement

Write the problems: 23 + 5 = ___ and 27 + 5 = ___. Tell your student that before adding, she can "look ahead" to see if the tens will be more. For 23 + 5, we can look at 3 + 5, know that it will be less than ten, and write down the tens of the answer (2). Then we can add the ones, and write that. For 27 + 5, we know that there will be one more ten, so we write down one more ten (3). Then add 7 and 5 (either by making the ten (10 + 2); or remembering that 7 + 5 = 12, and write down the ones (2).

Mental Math 13

Extra Practice, Unit 13, Exercise 3A, pp. 129-132

23 + 5 = ____
23 + 5 = 2____
23 + 5 = 28____
27 + 5 = ____
27 + 5 = 3____
27 + 5 = 32____

(7) Review: Subtract within 20 by subtracting from a ten

Activity

Write the expression: 13 – 8. Display 13 using base-10 materials and write 13 as a number bond with 10 and 3.

Ask your student if we can subtract 8 from the ones. There are not enough ones, so we cannot. Then ask if we can subtract 8 from the ten. We can. Have your student find the answer. Then ask her to tell you how many ones are left. Write the answer to the problem. You can add the number bond under 13 to show how the 13 was spit into tens and ones.

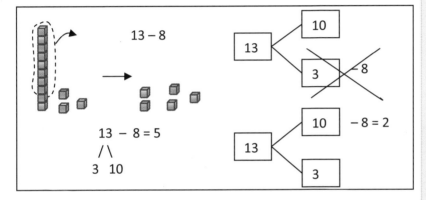

You can show your student another way to subtract 8 from 13. We can first subtract 3 to get down to the ten. Then we have to subtract another 5 from the ten (since 8 is 3 and 5).

Repeat with other examples, as needed.

Write the problems shown at the right and have your student find the missing number by subtracting from the ten.

$$17 - 9 = \underline{\quad} + 7$$
$$14 - 7 = \underline{\quad} + 4$$
$$11 - 6 = \underline{\quad} + 1$$

Workbook

Exercise 16, p. 58

Reinforcement

Use playing cards with face cards removed. Shuffle and place the deck face-down. Turn over the first card. This is the ones of a teen number; e.g. 5 is turned over and the number is 15. Write down the number 15. Turn over the next card. Have your student subtract the number turned over from the number written down. You can do this as a game by taking turns; the player with the smallest difference wins a point.

Mental Math 14

Extra Practice, Unit 13, Exercise 3A, pp. 129-132

Workbook

Exercise 16, p. 58

1.
$$9 \rightarrow 8$$
$$\downarrow$$
$$7 \leftarrow 8 \leftarrow 6$$
$$\downarrow$$
$$7 \rightarrow 8 \rightarrow 9$$
$$\downarrow$$
$$6 \leftarrow 9 \leftarrow 9$$
$$\downarrow$$
$$7 \rightarrow 8 \rightarrow 5$$

(8) Subtract ones from tens

Textbook

Tasks 6-7, pp. 42-43

6. (a) 19
   (b) 23
7. (a) 14
   (b) 22

Workbook

Exercise 17, pp. 59-60

1. 16   22
   33   11
   24   23
2. 21   15
   17   35
   13   32

Activity

Write the expression 10 – 8.

Ask your student for the answer.

Then write the expression: 30 – 8.

Illustrate with base-10 material. In order to remove 8, a ten needs to be traded in for 2 ones.

Show your student that we need to subtract 8 from one of the tens, which results in one less ten.

Show the process with number bonds as well.

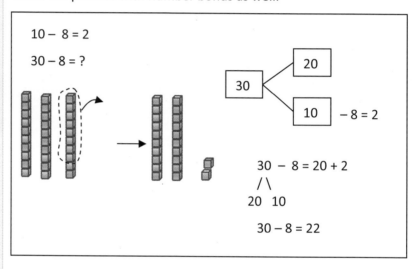

So 30 – 8 is 20 and 2, or 22.

Repeat with other examples.

Discussion

Tasks 6-7, (a) and (b) only, pp. 42-43

Workbook

Exercise 17, pp. 59-60

Reinforcement

Mental Math 15

(9) Subtract ones from a 2-digit number within 40 with renaming

Activity

Write the expression: 15 – 8.

Ask your student for the answer. Then write the expression: 35 – 8.

Illustrate with base-10 material. Show your student that since there are not enough ones to take away 8, we take it away from one of the tens. That leaves 2 ones. Ask your student how many are left. There are now 2 tens, 2 ones, and 5 ones, or 27.

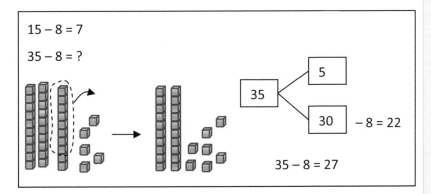

$15 - 8 = 7$

$35 - 8 = ?$

$35 - 8 = 27$

The process of subtraction can be thought of and shown in several ways:

⇒ Subtract 8 from 10. 10 – 8 = 2. There is one less 10.
  So 35 – 8 = 25 + 2 = 27.

$$35 - 8 = 25 + 2$$
$$/\backslash$$
$$25\ 10$$

⇒ Subtract 8 from 30. 30 – 8 = 22.
  So 35 – 8 = 5 + 22 = 27.

$$35 - 8 = 5 + 22$$
$$/\backslash$$
$$5\ 30$$

⇒ Subtract 8 from 15. 15 – 8 = 7. There is one less 10.
  So 35 – 8 = 20 + 7 = 27.

$$35 - 8 = 20 + 7$$
$$/\backslash$$
$$20\ 15$$

Provide additional problems, getting your student to Illustrate them with base-10 material and explain his thought process.

Textbook

Tasks 6-7, pp. 42-43

6. (c) 26
7. (c) 22

Workbook

Exercise 18, 61-62

1. 23

3; 33	2; 22
2; 22	2; 32
4; 34	2; 22

2. 14

7; 27	7; 17
8; 18	5; 15
8; 28	9; 29

Discussion

Tasks 6.(c) and 7.(c) pp. 42-43

Get your student to explain his thought processes in answering these.

Write some problems for your student to try to solve without manipulatives or pictures.

Workbook

Exercise 18, 61-62

Reinforcement

Give your student 4 dimes [10-cent coins] and some objects for him to buy, each tagged with 1 to 10 cents. Have him buy the objects, using pennies if he has enough, or a dime if he does not.

For example, tell him to buy something that costs 3 cents. Ask him how he would pay for it. To pay for it, he has to give a dime. Ask him how much change he should get. Give him the 7 cents change. Ask him how much money he has left. He has 37 cents.

Then, he buys something for 6 cents. He has enough pennies, and now has 31 cents.

Then he buys something for 5 cents. This time, he has to use a dime to pay for it and get change (subtract from a ten). See if he can tell you how much money he would have left before he buys the item.

Mental Math 16 and 17.

Extra Practice, Unit 13, Exercise 3B, pp. 133-136

Tests

Tests, Unit 13, 4A and 4B, pp. 45-47

Chapter 4 – Adding Three Numbers

Objectives

♦ Add three 1-digit numbers.

Material

♦ Multilink cubes

Notes

When we add three numbers, we can add each number in order, left to right. However, since numbers can be added in any order, they can be grouped in two other ways.

For example,

to add 8 + 4 + 2, we can add 8 + 4 =12, and then add 12 + 2 = 14.

Or, we can add 4 + 2 = 6, and then add 6 + 8 = 14.

Or, we can add 8 + 2 = 10, and then add 10 + 4 = 14.

$$8 + 4 + 2 = 12 + 2 = 14$$

$$8 + 4 + 2 = 6 + 8 = 14$$

$$8 + 4 + 2 = 10 + 4 = 14$$

This last method uses the strategy of making ten. Encourage your student to look for tens first when adding a series of numbers.

(1) Add three 1-digit numbers

Textbook

Page 44

10; 14

Tasks 1-2, p. 45

1. (a) 10
   (b) 15
2. (a) 12   (b) 13
   (c) 14   (d) 18
   (e) 16   (f) 16
   (g) 18   (h) 24

Discussion/Activity

Page 44

Have your student look at and discuss textbook p. 44. First there are 8 bags. The girl adds 2 more. Then the boy adds 4 more. We can find the total by first adding 2, then 4.

Illustrate this situation with linking cubes. Link together 8 cubes of one color, 4 cubes of another color, and 2 cubes of a third color.

Give your student the 8-rod and the 4-rod and ask her to add them together. Then give her the 2-rod and ask her to add that to the total. Write the addition equations, and then the equation showing 3 addends; 8 + 4 + 2 = 14.

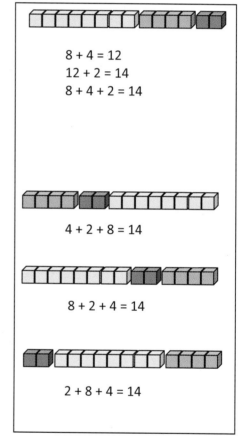

$$8 + 4 = 12$$
$$12 + 2 = 14$$
$$8 + 4 + 2 = 14$$

$$4 + 2 + 8 = 14$$

$$8 + 2 + 4 = 14$$

$$2 + 8 + 4 = 14$$

Show her that any arrangement of the 3 lengths end-to-end gives the same total length. Write the equations to reflect the different arrangements, and ask her to add the numbers in order.

Ask her if one order was easier to add with the other two. Lead her to see that it can be easier to add the numbers when the first two add to 10.

Workbook

Exercise 19, pp. 63-65

1. (a) 6
   (b) 9
2. 9   12
   20   10
3. (a) all 12
   (b) all 15

Write the expression 4 + 7 + 6 and ask your student to find the answer. Ask him how he added them. If he did not first add 4 and 6, point out that since the numbers can be added in any order, he should first look for a combination that makes 10.

Repeat with another example, such as 4 + 5 + 7. In this case, no pair of numbers make a ten. Your student might find it easier to start with the greatest number, 7, first, or he might find it easier to first add the smallest numbers together since he should have memorized 4 + 5 by now.

Discussion

Tasks 1-2, p. 45

Encourage your student to first see if any pair makes ten before adding.

Workbook

Exercise 19, pp. 63-65

Reinforcement

Game
Material: Number cards 1-9, 4 sets, or a deck of playing cards with face cards removed.
Procedure: Shuffle cards and deal all out. Players keep cards face-down. Each player turns over 3 cards and finds their total. The player with the greatest total wins a point.

Extra Practice, Unit 13, Exercise 4, pp. 137-138

Enrichment

Have your student add four or more numbers (keep the total within 40). Get him to first look for pairs that make 10. For example

$$5 + 2 + 6 + 5 + 4 = 22$$

Tests

Tests, Unit 13, 4A and 4B, pp. 45-47

Chapter 5 – Counting by 2's

Objectives

♦ Count forward by twos.
♦ Count backward by twos.

Material

♦ Multilink cubes

Notes

This chapter introduces counting by 2's using repeated addition. Students will first use concrete items and pictorial aid to count forwards and backwards by 2's. Then they will be count by 2's from memory, or from silently counting on. Eventually, your student should be able to "skip count" by 2's from memory.

Students have already learned to count by 10's. In unit 19, they will count by 5's within the context of counting nickels. In *Primary Mathematics 2*, they will learn how to count by 3's and 4's as well within the context of learning multiplication facts. In *Primary Mathematics 3*, they will count by 6's, 7's, 8's, and 9's.

Counting by 2's is generally taught by starting from 0, or, when counting backwards, starting from an even number. This allows students to become familiar with multiples of 2. However, in the context of repeated addition, we can count by 2's from any number, e.g. 1, 3, 5, 7, …

(1) Count by twos.

Activity

Give your student 20 multilink cubes and ask her to link them together into 2's. Take one 2-unit and set it in front of her and ask her how many there are. Add another 2-unit next to it, and ask her how many there are. Continue adding 2-units to the row, asking how many there are each time. She can add on to her previous answer to give you the total. Then, point to each 2-unit one at a time and count by 2's. Tell your student that you are counting by 2's. Normally we count by 1's, where the next number in counting is always 1 more. But we can also count by 2's, where the next number we say is 2 more. When we count by 2's, we are adding 2 each time.

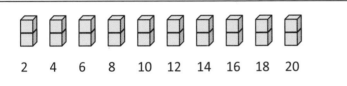

2	4	6	8	10	12	14	16	18	20

Get your student to practice counting by 2's by pointing to the 2-units one at a time and having him count. You can start by writing the numbers below the cubes and having him read them, then erasing some of the numbers and having him say the missing numbers as he counts by 2's, and then erasing all the numbers.

Repeat with 40 cubes linked in 2's.

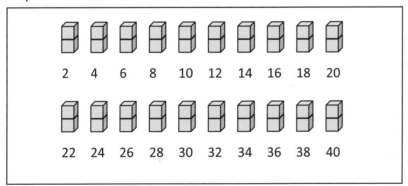

2	4	6	8	10	12	14	16	18	20
22	24	26	28	30	32	34	36	38	40

Write an addition expression involving just 2's and ask your student to find the answer by counting by 2's, pointing at each 2 in order to keep track.

$$2 + 2 + 2 + 2 + 2 + 2 = 12$$

Textbook

Page 46

Tasks 1-3, pp. 46-47

1. (a) 4
   (b) 6
   (c) 8
2. 12, 14; 18, 20; 28, 30
3. (b) 24, 26; 30; 34

Workbook

Exercise 20, pp. 66-67

1. 6, 8, 10, 12, 14, 16, 18
   18
2. 4, 6, 8, 10, 12
3.   4, 6, 8, 10
   12, 14, 16, 18, 20
   20
4. 22; 28, 30; 36, 38, 40

Make up some story problems that involve counting by 2's and ask your student to find the answer. For example, "A family had 5 children. When they had a new cement walkway poured for their new house, the parents let each child press both palms into the cement as it was drying to make palm prints. How many prints were there?"

Point at the last 2-unit and ask how many there are in all. (40), Move it aside, and ask how many there are now. (38), Continue until all the 2-units are gone. Tell your student that she is counting backwards by 2's. Replace the cubes, write the numbers under them, and have her practice counting by 2's backwards, eventually erasing all of the numbers.

Discussion

Page 46

Tasks 1-2, pp. 46-47

Reinforcement

Get your student to practice counting backwards and forwards by 2's to 20 or beyond without blocks periodically as you continue with the curriculum.

Give your student an even number within 30, such as 12, and ask your student to count on by 2's from that number to 40.

Extra Practice, Unit 13, Exercise 5, pp. 139-140

Enrichment

Have your student count by 2's from an odd number.

Tests

Tests, Unit 13, 5A and 5B, pp. 49-52

Review

Review

Review 8, pp. 68-73

Use the review in the workbook as an assessment to see if you need to re-teach any concepts.

Tests

Tests, Unit 13 Cumulative Tests A and B, pp. 53-58

Workbook

Review 8, pp. 68-73

1. 28   38
   29   19
2. 23, 24; 26, 27, 28; 30
   31, 32; 34, 35; 37, 38, 39, 40
3. 29, 32, 35, 37, 40
4. (a) 4
   (b) 28
   (c) B
   (d) C
5. (a) 8; 8   (b) 20; 20
   (c) 27; 27 (d) 40; 40
6. (a) 7
   (b) 5
   (c) 3
   (d) 4
   (e) guppies
   (f) swordtails
7. 14 − 6 = 8; 8
8. 11
9. 15
10. 2, 4, 6, 8, 10
11. 32
12. 28
13. (a) 6
    (b) 17
    (c) No
    (d) No
    (e) Bears
    (f) Dolls

Unit 14 – Multiplication

Chapter 1 – Adding Equal Groups

Objectives

♦ Understand the concept of equal groups.
♦ Find the total number of objects in a given number of equal groups.
♦ Use mathematical language such as "3 fives" or "3 groups of 5" to describe equal groups.
♦ Find the total number in equal groups by repeated addition.

Material

♦ Counters
♦ Multilink cubes
♦ Paper plates or bowls (for groups)

Notes

Addition and subtraction can be interpreted in terms of part and whole. When a whole is made up of two parts, we add to find the whole given the two parts. We subtract to find one part given the whole and the other part.

The part-whole concepts of addition and subtraction can be extended to multiplication and division when the whole is made up of multiple equal parts. We multiply to find the whole given the number of parts and the number in each part. We divide to find the number in each part given the whole and the number of parts. We also divide to find the number of parts given the whole and the number in each part.

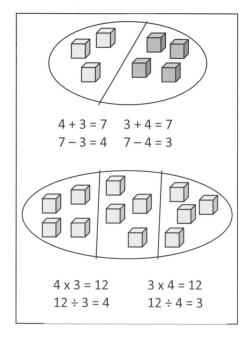

$4 + 3 = 7$ $3 + 4 = 7$
$7 - 3 = 4$ $7 - 4 = 3$

$4 \times 3 = 12$ $3 \times 4 = 12$
$12 \div 3 = 4$ $12 \div 4 = 3$

Multiplication means putting together equal groups. In this section, students will learn to recognize equal groups. They will count the number of groups and the number in each group. Then they will find the total number in the groups by repeated addition. In this unit, multiplication and division will be within 40. Students should use mental math to do the repeated addition, not simply count on. For example, they should be able to add $4 + 4 + 4 + 4$ by thinking "4 and 4 is 8, 8 and 4 is 12, 12 and 4 is 16" or simply "4, 8, 12, 16" adding 4 mentally each time.

Students will not be memorizing multiplication facts until *Primary Mathematics 2*. They learned "skip counting" by 2's in the last chapter. If you want to teach them "skip counting" by 3 and 4 through memorization, you can, but that will be taught in *Primary Mathematics 2*, and continued practice in mental math through repeated addition is beneficial at this stage. Skip counting by 5 is easy to learn, though, and will be covered in unit 19 since it is helpful in counting coins.

(1) Put the same amount in each group and find the total

Activity

This lesson focuses on the number in each group.

Use up to 40 multilink cubes or other counters. Set out some bowls, such as 3 bowls. Ask your student to put a specified number of counters in each bowl. Ask him

⇒ How many are there in each group?
⇒ How many are there altogether?

Encourage your student to find the total by adding mentally, if he can, rather than by simply counting. If it makes it easier, write down his total each time, e.g. 5, 10, 15. 4, 8, 12, etc. Note that your student is not memorizing how to skip count yet, but rather using mental addition each time.

Write something similar to the following (for 3 groups of 5) , read them aloud using "how many" or "what" for the blanks, and ask your student to say and write the answer,

⇒ There are ___ counters in each group.
⇒ 5 + 5 + 5 = ___

Repeat with other examples. Emphasize that there are *equal* groups because there is the same amount in each group.

Discussion

Page 48-49

Discuss the arrangement of the fruit, guiding your student to see that the fruit are arranged in equal groups. Ask questions such as

⇒ How many types of fruit are there?
⇒ How many plates of pears are there?
⇒ How many pears are on each plate?
⇒ Does each plate have the same number of pears?
⇒ How many pears are there altogether?

Tasks 1-3, p. 50

Workbook

Exercise 2, pp. 74-77

Reinforcement

Look for opportunities to discuss multiplication in the environment. For example, there are 5 chairs. Each chair has 4 legs. There are 5 fours. How many legs are there?

Textbook

Page 48

Tasks 1-3, p. 50

1. 2, 12
2. 8, 24
3. 5, 20

Workbook

Exercise 1, pp. 74-75

1. 6; 6
   12; 12
   12; 12
   12; 12
2. 6
   8
   15
   8

Exercise 2, pp. 76-77

1. (a) 5; 10
   (b) 2; 8
   (c) 10; 30
2. (a) 6
   (b) 12
   (c) 16
   (d) 10

(2) Make groups of the same amount and find the total

Textbook

Page 49

    15; 15
    24; 24
    12; 12

Tasks 4-6, p. 51

  4. 4, 12
  5. 5, 20
  6. 3, 7, 21

Workbook

Exercise 3, pp. 78-79

  1. (a) 4; 20
     (b) 3; 18
     (c) 4; 20
     (d) 3; 18
  2. (a) 15
     (b) 12
     (c) 8
     (d) 10

Activity

This lesson focuses on the number of groups.

Use counters and multilink cubes and bowls again. This time give him some counters, such as 20 of them. He should not count them first. Tell him to make equal groups of 5 counters. He need to put 5 in a group, then five more.

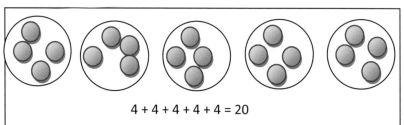

$$4 + 4 + 4 + 4 + 4 = 20$$

Ask him how many groups he has. Then ask him to tell you how many counters there are, again using repeated addition. Write the problem as an addition problem. Encourage him to add mentally. He can point to the second 5, say 10, point to the next 5 and add 5 mentally.

Repeat with other examples.

Write a few problems of repeated addition and see if your student can find the answer without counters using mental addition. The total should be within 40.

Discussion

Tasks 4-6, p. 51

Workbook

Exercise 3, pp. 78-79

Reinforcement

Make patterns with equal groups, such as threading a string with equal groups of different colors of beads. Or, if your student likes to draw, get her to draw pictures where the items are in equal groups, such as 4 apple trees with the same number of apples on each tree, or ponds with fish, or rows of tanks. Discuss the number of groups, the number in each group, and the total.

Extra Practice, Unit 14, Exercise 1, pp. 143-144

Tests

Tests, Unit 14, 1A and 1B, pp. 59-62

Chapter 2 – Making Multiplication Stories

Objectives

◆ Write multiplication equations using "x" and "=" for a given situation involving multiplication.
◆ Interpret multiplication equations with manipulatives or illustrations.

Material

◆ Counters
◆ Multilink cubes

In this chapter the term "multiplication" and the multiplication sign "x" are introduced.

At this level, we will interpret 3 x 5 as "3 groups of 5" or "3 fives". In *Primary Mathematics 2*, students will learn that 3 x 5 = 5 x 3 so they could also write the multiplication expression as 5 x 3, or 5 each in 3 groups.

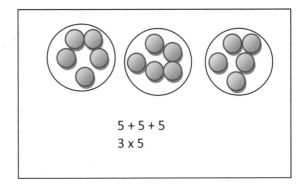

5 + 5 + 5
3 x 5

⇒ Students should interpret 3 x 5 as
⇒ 3 fives
⇒ 3 groups of 5
⇒ Multiply 3 and 5

At this level, the emphasis is on understanding the meaning of multiplication rather than on memorizing the multiplication facts. Encourage your student to use manipulatives or drawings to illustrate the meaning of multiplication, and to use repeated addition to find the answer. Do not memorization of multiplication facts yet.

You may want to occasionally read multiplication expressions as ___ *of* ___. For 3 x 5, there are "three of groups *of* 5" or 3 *of* the fives". Even though the wording is awkward, this might help reduce confusion later when the multiplication symbol is used for a fraction of a whole, e.g. $\frac{1}{4}$ of 8 is $\frac{1}{4}$ x 8.

(1) Write multiplication equations

Textbook

Pages 52-53

Tasks 1-2, pp. 54-55

Workbook

Exercise 4, pp. 80-81

1.	5 x 4	3 x 8
	3 x 8	5 x 4
	5 x 4	3 x 8
	6 x 3	4 x 10
	4 x 10	6 x 3
	6 x 3	4 x 10
2.	2 x 4	4 x 5
	3 x 3	2 x 5
	5 x 2	3 x 4

Discussion

Pages 52-53

Get your student to use mathematical language to describe the blocks on the plates, e.g.:

⇒ 4 groups of 2
⇒ 4 twos

Tell your student that we can also read "x", the multiplication symbol, as *times*. 4 x 2 = 8 can be read as "4 *times* 2 equals 8". We have 2, four times.

Point out the two expressions, 2 + 2 + 2 + 2 and 4 x 2. Tell your student that they both mean the same thing.

Ask your student which math equation is easier to write;

⇒ 2 + 2 + 2 + 2 = 8
⇒ 4 x 2 = 8.

Tasks 1-2, pp. 54-55

Emphasize that multiplication is used when we have equal groups. For task 2, point out that it is the number of legs that are being counted.

Use counters or other objects to make equal groups, or discuss situations in the environment, such as 5 chairs with 4 legs. Get your student to write a multiplication expression for the situation, e.g. 5 x 4.

Workbook

Exercise 4, pp. 80-81

Reinforcement

Ask your student to write multiplication equations for Exercises 1-3 in the workbook.

(2) Interpret multiplication equations

Activity

Use counters or multilink cubes. Write a multiplication expression, such as 2 x 8. Ask your student to illustrate the expression with the counters or cubes, i.e., make 2 groups of 8 counters, or link 8 cubes together twice. Ask your student to tell you the total and write the completed equation, 2 x 8 = 16.

Repeat with other examples. Students can also draw circles with X's or other symbols in them to illustrate the expressions.

Discussion

Task 2, p. 55

Workbook

Exercise 5, p. 82

Reinforcement

If your student likes to draw pictures, write several multiplication expressions and get him to draw one picture where each expression is illustrated. For example, write 2 x 3, 4 x 5, 8 x 2. Your She could draw a picture with two tricycles, four apple trees each with 5 apples on them, 8 birds flying. Ask her to explain which are the groups, and which are the items in the group. For example: There are two tricycles. Each tricycle has 3 wheels.

Extra Practice, Unit 14, Exercise 2, pp. 145-146

Tests

Tests, Unit 14, 2A and 2B, pp. 63-69

Workbook

Exercise 5, p. 82

Check drawings

Chapter 3 – Multiplication Within 40

Objectives

♦ Multiply within 40 using repeated addition.
♦ Use rectangular arrays to illustrate multiplication.
♦ Solve picture problems involving multiplication.

Material

♦ Counters
♦ Multilink cubes

Notes

In this chapter, picture problems are used to reinforce the concept of multiplication as repeated addition. Students should recognize that the groups are equal and count to find the number of groups and the number in each group. If your student has difficulty adding with mental math, allow him to find the answer by counting, but continue practice with mental math.

The problems illustrate real-life situations involving groups and rectangular arrays. Teach your student the terms "column" and "row." She can think of either the columns or the rows as the groups.

(1) Multiply within 40 using repeated addition

Activity

Continue to provide practice with multiplication situations:

Display equal groups and get your student to write the multiplication equation and find the answer.

Write a multiplication expression and have your student illustrate it with pictures, toys, counters, or multilink cubes and find the answer.

Give your student up to 40 counters and have her form equal groups and write the multiplication equations.

Tell your student a story that involves multiplication, and get him to write a multiplication equation and work out the answer with manipulatives.

Write a multiplication expression and ask your student to tell a story or give a situation where the expression would be used.

Discussion

Page 56

Tasks 1-3, pp. 57-58

Workbook

Exercise 6, pp. 83-85

Reinforcement

Write the following number patterns and ask your student which number is next in the pattern.

⇒ 2, 4, 6, 8, ____

⇒ 3, 6, 9, 12, ____

⇒ 4, 8, 12, 16, ____

⇒ 5, 10, 15, 20, ____

⇒ 6, 12, 18, 24, ____

Textbook

Page 56

  12; 20

Tasks 1-3, pp. 57-58

1. 6; 6
2. 20; 20
3. (a) $2 \times 5 = 10$
   (b) $5 \times 2 = 10$
   (c) $6 \times 3 = 18$

Workbook

Exercise 6, pp. 83-85

1. 4×3
   3×2
   5×2
   2×6
   5×3
   4×5
2. (a) 6
   (b) 12
   (c) 20
   (d) 15
3. (a) 12
   (b) 18
   (c) 12
   (d) 24
   (e) 14

(2) Write multiplication equations for rectangular arrays

Textbook

Tasks 4-5, p. 59

  4. 18; 18
  5. 20; 20

Workbook

Exercise 7, pp. 86-88

  1. 3 x 3 = 9; 9
  2. 4 x 3 = 12; 12
  3. 5 x 3 = 15
     or 3 x 5 = 15; 15
  4. 8 x 3 = 24
     or 3 x 8 = 24; 24
  5. (a) 3 x 5 = 15
     (b) 5 x 2 = 10
     (c) 4 x 3 = 12
     (d) 6 x 3 = 18

Activity

Use counters to create 3 rows of 4. Put them on paper or a flat whiteboard so you can draw around them.

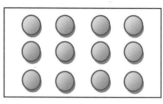

Tell your student that this arrangement is called is made up of rows and columns. Point out the rows and columns. Ask her how many counters there are in each row, and how many in each column. Discuss other instances where your she has seen objects arranged in an array, such as cookies on a cookie sheet, the key pad on a phone or a keyboard.

Draw boxes around each column. Tell your student that each *column* is a group. Ask him how many groups there are, and how many in each group. There are 4 columns of 3. Write the multiplication equation.

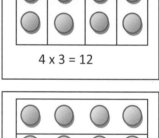

4 x 3 = 12

Erase the boxes, and draw new boxes around each row, or make another array of 3 rows of 4. Tell your student that each *row* is now a group. Ask her how many groups there are, and how many in each group. There are 3 rows of 4. Write the multiplication equation.

3 x 4 = 12

Use multilink cubes. Write a multiplication expression and have your student form an array using the cubes to represent the expression. Then have him find the total. He can have the groups be the rows or the columns.

Discussion

Tasks 4-5, p. 59

Workbook

Exercise 7, pp. 86-88

Reinforcement

Extra Practice, Unit 14, Exercise 3, pp. 147-148

Tests

Tests, Unit 14, 3A and 3B, pp. 71-77

Review

Review

The workbook has 3 reviews in a row. You may want to do the first one now, the next one after Unit 15, adding in a few division situations, and the third one after Unit 16. You can also do one page of review with each lesson for more continuous review. Reviews include concepts from *Primary Mathematics* 1A.

Reinforcement

Use a number cube labeled with 1-6 and multilink cubes. Have your student throw the number cube twice. The first number is the number in each group. Get him to link that many cubes in a row. The second number is the number of groups. Get him to add rows to his first row to make an array based on the two numbers. Ask him to find the total number of cubes and write a multiplication equation.

For example, he first throws a 5 and then a 6, so he links 5 cubes and then adds more rows of 5 until he has 6 of them, so that he has a 6 by 5 array. Then he writes 5 x 6 = 30 (or 6 x 5 = 30). Repeat for more practice.

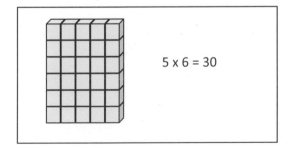

5 x 6 = 30

Tests

Tests, Unit 14 Cumulative Tests A and B, pp. 79-86

Workbook

Review 9, pp. 89-92

1. (a) 33   (b) 21   (c) 30   (d) 32
2. 6
3. (a) 20   (b) 12   (c) 21   (d) 24
4. (a) 20   (b) 4   (c) apple
   (d) orange   (e) 2   (f) 4
5. 10 – 6 = 4; 4
6. 5 + 7 = 12; 12
7. 17 – 5 = 12; 12

Review 10, pp. 93-96

1. (a) 10; 20; 30; 40   (b) 3; 13; 23; 33
   (c) 18; 28; 38   (d) 29; 19; 9
2. (a) Dan; 4   (b) 2
3. (a) 9   (b) 3
4. (a) 27, 28; 30, 31   (b) 10, 12, 14, 16
   (c) 20, 25; 35, 40
5. (a) 36   (b) 32   (c) 38   (d) 26
6. (a) 14   (b) 10   (c) 7   (d) 18
7. 19 – 6 = 13; 13
8. 9 – 4 = 5; 5
9. 5 + 4 + 2 = 11; 11

Review 11, pp. 97-100

1. (a) 34; 14; 20; 34   (b) 15
2. e.g.
3. check answers
4. (a) 8   (b) 6
5. C, A, B. D
6. mango
7. 12; 12
8. 11; 11
9. 3; 3
10. 11 – 9 = 2; 2
11. 8 = 5 = 3; 3
12. 12 – 8 = 4; 4

Unit 15 – Division

Chapter 1 – Sharing and Grouping

Objectives

♦ Understand division as sharing.
♦ Use manipulatives to illustrate sharing.
♦ Understand division as grouping.
♦ Use manipulatives to illustrate grouping.

Material

♦ Counters
♦ Multilink cubes
♦ Bowls

Notes

Division is the opposite of multiplication. In multiplication, we put together equal groups. In division we break up into equal groups. There are two types of division situations.

Sharing
    Start with a set of objects. Make a given number of equal groups. Then find the number in each group. For example, share 12 cookies equally among 3 children. 12 is the total, 3 is the given number of groups. The division can be accomplished by first giving each child 1 cookie, then giving each child another cookie, and so on until all are distributed. Each child gets 4 cookies. 4 is what we need to find; the number in each group. Sharing is illustrated on p. 55 of the textbook.

Grouping
    Start with a set of objects. Make equal groups of a given size. Then find the number of groups made. For example, put 12 marbles into groups of 3. 12 is the total, 3 is the number that goes in each group. The division can be accomplished by first putting 3 in one group, then 3 in another, and so on. Then find the number of groups. There are 4 groups. Grouping is illustrated on p. 56 in the textbook.

    Both of the situations described above have the same division equation; $12 \div 3 = 4$. Therefore, $12 \div 3 = 4$ can be interpreted as either grouping or sharing.

    At this level, students will not be required to write division equations. The division symbol will be introduced in *Primary Mathematics* 2.

Sharing

$$12 \xrightarrow[(\div 3)]{\text{Shared among 3}} 4$$

Grouping

$$12 \xrightarrow[(\div 3)]{\text{Make groups of 3}} 4$$

$12 \div 3 = 4$

(1) Divide by sharing

Activity

Use counters and bowls. Give your student 20 counters and 4 bowls. Ask her to divide the counters up equally among the bowls. Let her do it any way she wants – she may already know to put 5 in each bowl, or she may start by putting 1 in each bowl, then another in each bowl, and so on. Ask her how many there are in each bowl and if there is the same number in each bowl.

Tell your student that she divided up the counters into equal groups. With multiplication, we know the number of groups and how many are in each group. Division is the opposite – we have a total and want to make equal groups.

Discuss situations where we might want to divide to make equal groups, such as sharing cookies.

Discussion

Top of page 60 and page 61

Tasks 1-3, pp. 63-64

You can have your student act out these situations using counters so that he becomes familiar with the process of sharing, since the pictures in the textbook give only the final result. For example, in task 1, give your student ten counters and ask him to put them into 2 groups before looking at the picture.

Workbook

Exercise 1, pp. 101-104

Note: Since Exercise 2 has one division by grouping problem, assign it after the next lesson rather than now.

Reinforcement

Give your student some counters or multilink cubes or other objects, up to 40, and tell him the number of groups you want him to put them in. Ask him to find the number that goes in each group. Avoid situations where there is a remainder.

Extension

Include situations where there is a remainder. Tell your student that not all numbers divide evenly. Explore situations where there are and are not remainders. For example, 20 can be divided evenly by 4 or 5, but not by 3.

Textbook

Pages 60-61

5

Tasks 1-3, pp. 63-64

1. 5
2. 6
3. 7

Workbook

Exercise 1, pp. 101-104

1. (a) 3; 4
   (b) 4; 2
   (c) 2; 10
2. (a) 5
   (b) 3
   (c) 3
3. 6
4. 7
5. 3
6. 8
7. 4
8. 5

(2) Divide by grouping

Textbook

Pages 60, 62

Tasks 4-6, pp. 64-65

4. 2
5. 4
6. 4

Workbook

Exercise 2, pp. 105-106

1. 5
2. 3
3. 4
4. 5

Exercise 3, pp. 107-108

1. (a) 5
   (b) 6
   (c) 6
2. (a) 5
   (b) 6

Activity

Use multilink cubes. Give your student 20 cubes. Ask him to make groups of 4. He can link 4 together to make each group. Then ask him how many groups he has.

Tell your student that what he is doing is also dividing, but this time he needs to find the number of groups, rather than the amount that goes in the groups.

Discuss situations where we might want to find the number of groups, such as making cookies for a bake sale, putting 3 in each baggie, and wanting to know how many baggies are needed.

Discussion

Bottom of page 60 and page 62

Tasks 4-6, pp. 64-65

You can have your student act out the tasks using counters, so that she becomes familiar with the physical process of grouping. The textbook only shows the final result.

Workbook

Exercises 2-3, pp. 105-108

Note: You may want to assign Exercise 2 after Exercise 3, since it has 3 division by sharing problems and 1 division by grouping problems and thus could be used as a review.

Reinforcement

Give your student up to 40 cubes or other objects, and ask her to make groups of a given number, and then tell you the number of groups. There should not be any remainders. Mix in some sharing problems as well.

Extra Practice, Unit 15, Exercise 1, pp. 151-152

Tests

Tests, Unit 14, 3A and 3B, pp. 71-77
Tests, Unit 14 Cumulative Tests A and B, pp. 79-86

Enrichment

Give your student 11 cubes and ask her to try to divide them into equal groups, either by sharing or grouping. Then tell her there are some numbers with which you cannot make equal groups. Let her experiment with all the numbers from 1 to 20 to see which cannot make equal groups. She can also explore different ways that some numbers can make equal groups.

Unit 16 – Halves and Fourths

Chapter 1 – Making Halves and Fourths

Objectives

♦ Recognize halves and fourths.

Material

♦ Paper squares, rectangles, and circles
♦ Paper strips
♦ Food that can be divided, such as apple, large cookie, pan of brownies.
♦ Appendix p. a13

Notes

This unit on fractions deliberately follows the unit on division. Your student will look at another type of dividing into equal parts, this time to get a fraction of a whole. In a later level, he will eventually formally relate fractions to division.

The words **half** and **quarter** are used to describe a part of a whole. One **half** of a whole is one of two **equal parts** that make up the whole. A **fourth** of a whole is one of four **equal parts** which make up a whole. A fourth is sometimes called a quarter. Use both terms, fourths and quarters, with your student.

Your student will learn the concepts of half and fourth informally through paper folding and cutting an object. Food is an easy object to use to illustrate fractions since your student is used to sharing or cutting up food into parts.

Fractional notation such as $\frac{1}{2}$ and $\frac{1}{4}$ will be introduced in *Primary Mathematics* 2.

Students were introduced to patterns in Kindergarten and recognize extended patterns based on shape, size, and orientation of the whole shape. Here, they will recognize and extend patterns based on the orientation of a half or a fourth of the shape.

(1) Recognize halves and fourths

Textbook

Page 66

Tasks 1-3, p. 67

1. 2, 4
2. (a), (c)
3. (a), (b)

Activity

Your student is probably familiar with the concept of a half. Tell him you want to share an apple or cookie with him. Cut one into half, as evenly as possible. Cut another apple or cookie into two unequal parts. Ask him which cut resulted in two equal parts. Tell your student that when the parts are equal, each part is **one half of the whole**.

Show your student a paper circle. You can tell her it represents a cake or a pizza. Fold it in half so that the two folded edges match. Unfold it and draw a line on the crease. Tell her that this divides the circle into halves. Cut along the crease. Place the two pieces together to show the original circle, and then on top of each other to show that they are equal. Tell her that when we make two *equal* parts from one whole, each part is called a **half**.

Use paper squares or rectangles or index cards and draw lines on them, sometimes showing half and sometimes not. Ask your student if one part is a half. You can draw other shapes as well, such as triangles, and ask him how they could be cut in half.

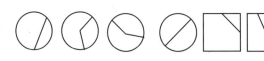

Workbook

Exercise 1, pp. 109-110

1. (a) yes
    (b) no
    (c) yes
    (d) no
    (e) yes
    (f) no
    (g) no
    (h) yes

Exercise 2, pp. 111-112

    Check answers

Give your student some square pieces of paper. Ask him to fold it and cut it into two halves in various ways.

Show your student a paper circle. Fold it in half and then half again. Draw lines at the creases. Ask her to count the parts and tell you if the parts are equal. Tell her that each part is a **fourth**, or a **quarter**, of a circle. Cut the parts out to show that each part is the same.

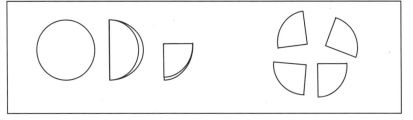

Give your student some square pieces of paper. Ask her to fold them and cut them into fourths in different ways. Help her find other ways to make fourths.

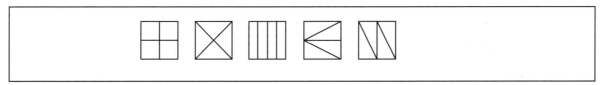

Discussion

Page 66

Tasks 1-3, p. 67

Reinforcement

Have your student divide other shapes, such as hexagons, triangles, or rectangles, into halves or fourths.

Get your student to find out which letters of the alphabet can be divided into equal halves or fourths. See the appendix page a13 for letters. You can include in the discussion ways that the letters could be drawn so that they could be divided into halves, e.g., both parts of the B could be drawn equally, or the line on the Q could be drawn straight, and the letter be divided through the line.

Extra Practice, Unit 16, Exercise 1, pp. 155-156

(2) Recognize patterns

Activity

Review pattern recognition. Use different shapes or colors or other items (e.g. toy cars and planes) to create an ABAB pattern and then an AABAAB pattern. Get your student to say the pattern out loud, and then tell you what would come next. For example, if using linking cubes, the pattern can be red, blue, red, blue,... or red, red, blue, red, red, blue, ... Then do some patterns where the orientation of the objects changes, such as that shown here for triangles.

Workbook/Discussion

Exercise 3, pp. 113-114

Discuss the pattern in 1(a). See if your student can recognize that the shaded area flips relative to the center line, first from left to right, then back to left. If your student has trouble with this, trace and cut out the outer shapes of the figures, then trace and cut out the shaded portion. Show how the shaded potion flips to match each of the pictures. Get your student to say the pattern, "left, right, left, right, ...", to determine what comes next.

You might also need to discuss 1(d) or 2(d) where the shaded area is rotated. You can trace and cut out the figures, shading a portion, and show how shape rotates to fit each step of the pattern. Your student can say "top left, top right, bottom right, bottom left..." if needed to determine the pattern and what comes next.

Reinforcement

Get your student to create patterns with counters, coins, multilink cubes, pattern blocks, etc. for you to continue.

Tests

Tests, Unit 16, 1A and 1B, pp. 99-102
Tests, Unit 16 Cumulative Tests A and B, pp. 103-109

Workbook

Exercise 3, pp. 113-114

1. (a) right half
   (b) bottom
   (c) left
   (d) top right
2. (a) bottom
   (b) bottom right
   (c) bottom right
   (d) middle

Unit 17 – Time

Chapter 1 – Telling Time

Objectives

- Tell time to the hour and half hour on an analog clock face.
- Write times for the hour and half hour (e.g. "6 o'clock" or "half past 4").
- Relate time to the hour and half hour to events of the day.
- Relate time to another time or to an even using before or after.

Material

- Demonstration clock with geared hands, or real analog clock.

Notes

In this unit, students will learn to tell times on the hour and half past the hour. For most students, this is a review of what is learned in Kindergarten.

Although most clocks now are digital, students should be acquainted with the standard clock face (analog clock) and how the long hand and the short hand show the time and move relative to each other. They will have a better understanding of time and time estimation if they can "visualize" an hour, and later the minutes, as an amount of turning around the clock face. Terms such as "clockwise" are still in use today. Understanding how the hands turn relative to each can help the student understand angles as the degree of turning in later grades.

You can use the terms long hand and short hand rather than hour hand and minute hand, but if your student already has some understanding of minutes and hours go ahead and use the terms hour hand and minute hand. Students will learn to tell time on the minute in *Primary Mathematics* 2.

At 2 o'clock, the long hand points to the number 12 while the short hand points to the number 2.

From 2 o'clock to half past 2, the long hand turns through half of the clock face to point to the number 6, and the short hand moves half way from 2 to 3.

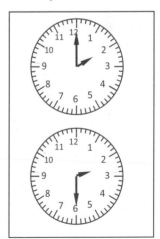

Students should also be able to sequence daily events with respect to clock time.

Your student has probably learned to tell time to the hour and half-hour in Kindergarten. You may be able to combine lessons for this unit.

(1) Tell time on the hour and sequence events according to time

Activity

Discuss with your student familiar uses of time and activities that you do on the hour, such as, "At 7 o'clock it is time to get up," or "We have lunch at one o'clock." Tell her that we divide time into hours. There are 24 hours in a day. Help her get an idea of how long an hour is. Select a familiar activity that takes an hour and say, "It takes an hour to...." Ask her to estimate other activities for whether they take less than an hour or longer than an hour, such as, "Does it take an hour to brush your teeth?"

Show your student various clocks that help us tell what time of day it is (without reading the time).

Show your student an analog clock. Point out the numbers and tell him that these mark the hours. Show him how the hands move around the clock from one hour to the next, and how the short hand points to the hour when the long hand is on 12, and that as the long hand goes around from 12 back to 12, the short hand moves to the next hour. Tell him that normally the time it takes for the long hand to go from 12 to 12 is one hour (on a real clock without pushing the hand around faster than normal). Explain that the hands always move in the same direction, a "clockwise" direction. Explain that **o'clock** means on the clock.

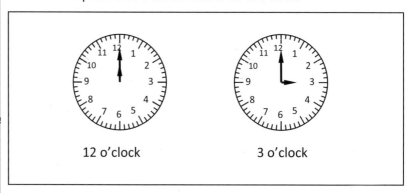

12 o'clock   3 o'clock

Set the clock for 12 o'clock and discuss what goes on at 12 midnight (it is night time, children are asleep). Continue with each hour through the day, showing that at noon the hours start over again. There are 12 hours on the clock, and the clock goes through the 12 hours twice in a day. So there is a 7 o'clock in both the morning and the evening, for example.

Get your student to move the hands on the clock and see how they move relative to each other. Ask her to set various times on the hour.

Textbook

Page 68

Task 1, p. 69

1. 7 o'clock
   9 o'clock
   11 o'clock
   2 o'clock
   6 o'clock
   9 o'clock

Workbook

Exercise 1, pp. 115-117

1. 1 o'clock   8 o'clock
   3 o'clock   5 o'clock
   7 o'clock   10 o'clock
   9 o'clock   11 o'clock
2. (a) 1 o'clock or 1:00
   (b) 3 o'clock or 3:00
   (c) 5 o'clock or 5:00
   (d) 7 o'clock or 7:00
   (e) 8 o'clock or 8:00
   (f) 10 o'clock or 10:00

You may want to tell your student that hours are divided up into minutes, and there are 60 minutes in the day. The long hand on a face clock points to the number of minutes, or how far into the hour it is before the next hour. So the 6:00 means the time is right at that hour.

Discussion

Page 68

Discuss the various time-pieces and ways to tell time. There is an analog clock, a digital clock, and a watch. There is also the sun, which does not tell us the exact time of day, but from which we can get an approximate time. Tell your student that 6 o'clock is the same as 6:00, and when it is exactly 6:00 o'clock, a digital clock (one without a face and hands, such as the one on the bedside table) will show 6:00.

Task 1 , p. 69

Workbook

Exercise 1, pp. 115-117

(2) Tell time on the half hour

Textbook

Tasks 2-3, pp. 70-71

2. 5
    7
    11
3. half past 9   10 o'clock
    half past 11   12 o'clock
    1 o'clock   half past 1

Workbook

Exercise 2, pp. 118-120

1. half past 2   half past 6
    5 o'clock   half past 10
    6 o'clock   half past 7
2. 9 o'clock or 9:00
    half past 9 or 9:30
    half past 12 or 12:30
    half past 2 or 2:30
3. 3 o'clock or 3:00
    half past 1 or 1:30
    half past 11 or 11:30
    half past 8 or 8:30
    half past 4 or 4:30
    12 o'clock or 12:00

Activity

Show your student a clock and draw an imaginary line from the 12 to the 6. You can also to show this with a drawn circle where the hours are marked. Ask her what each equal part of the circle is called. Each part is half of the whole.

Show your student a time on the hour, such as 2 o'clock. Move the hour hand half way around. Ask him to describe to you how far the hand move around the clock and the position of both hands. Guide him to say that the long hand has moved half way around, and the short hand has moved halfway from 2 to 3. Tell him that when the long hand points to 6, and the short hand is half-way between 2 and 3, we say that the time is **half past 2**.

Repeat with some other times on the hour and half past the hour.

Discuss some activities that would take about half of an hour. Ask your student whether other activities might take more than half an hour or less than half an hour.

2 o'clock

half past 2

Set some times to half-past the hour and get your student to tell you the time. Ask her to set some times to the half hour. For example, ask her to show you half past 6.

Discussion

Tasks 2-3, pp. 70-71

For p. 70, point out that when a face clock shows half past the hour, a digital clock shows the hour followed by :30 instead of :00. So half past 4 is the same as 4:30. You may want to tell your student that this is because the :30 means 30 minutes after the hour, and 30 is half of 60, the number of minutes in an hour.

Workbook

Exercise 2, pp. 118-120

Reinforcement

Relate routine activities of the day to the approximate time (hour or half past the hour).

(3) Relate times to other times, before and after

Activity

Set a time on the clock or look at a real clock. It does not have to be on the hour or half-hour. Ask your student if it is before or after a certain time. Use terms such as "in the morning" or in the evening". For example, it is morning, and the clock shows 11:20. Ask her questions such as, "Is it before or after 11:00 o'clock? Is it before or after 11:30? Is it before 10:00 in the morning? Is it before or after 12:00 noon? Is it before or after 1:00 in the afternoon?" Keep the discussion to the same day.

Set a time on the clock, such as 2:00. Discuss what could be happening at that time, being sure you discuss both 2:00 o'clock in the morning and 2:00 o'clock in the afternoon. Then say, "Pretend it is now 2:00 in the afternoon." Ask your student whether he normally does some activities before or after this time. For example, "Do you get up this morning before this time?", "Are we having this math lesson before or after this time?"

Discussion

Task 4, p. 72

Workbook

Exercise 3, pp. 121-122

Reinforcement

Extra Practice, **Unit 17, Exercise 1A and 1B, pp. 159-162**

Tests

Tests, **Unit 17, 1A and 1B, pp. 111-114**

Textbook

Task 4, p. 72

  4. (a) yes
     (b) no

Workbook

Exercise 3, pp. 121-122

  1. 1st picture → 11:00
    2nd picture → 2:30
    3rd picture → 530
  2. half past 1 or 1:30
    No
    Yes

Chapter 2 – Estimating Time

Objectives

♦ Estimate time to the nearest half-hour.
♦ Compare the time it takes to do specific activities.

Material

♦ Demonstration clock with geared hands, or real analog clock

Notes

In this chapter, students will be able to recognize time on a clock face that is almost to the hour or half-hour, or just past the hour or half-hour. Clock faces hardly ever show time exactly on the hour or half-hour. If your student can estimate the time to the closest half hour, she can use a clock to estimate time before learning how to tell time to the minute.

In this chapter, students will also determine whether one activity takes longer than another. This gives them a better understanding of the passing of time. Your student probably already knows when one activity takes longer than another. Sometimes, though, it might seem like time is passing more slowly or an activity that is boring takes longer than one that is more interesting. So some time estimates might be subjective. Also, the activities shown in the textbook are not very precise. It could take longer to get ready for bed than to have a story read depending on how long the child dawdles. In discussing time estimates, allow for such variation from the expected answers.

(1) Estimate time to the nearest half-hour

Activity

Show your student a time that is just before the hour. Tell her that the time is "almost" the hour, e.g., almost 2 o'clock. Point out that the long hand is almost to 12 and the short hand is almost to 2. Then show her a time that is just after the hour. Tell her that the time is "just after" or "just past" the hour, e.g. just past 2 o'clock. Point out that the long had is a bit past 12, and the short hand is a bit past 2.

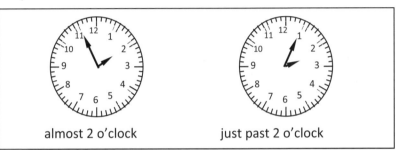

almost 2 o'clock   just past 2 o'clock

Repeat with times that are just before or just after the half hour.

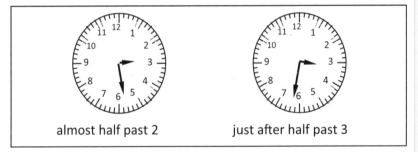

almost half past 2   just after half past 3

Discussion

Page 73, Tasks 1-2, pp. 74-75

Workbook

Exercise 4, pp. ~~45-47~~ 123-124

Reinforcement

Have your student look at the clock at various times of the day and help him to recognize a time during the first quarter hour as past the hour, during the second quarter hour as before the half hour, during the third quarter hour as after the half hour, and during the fourth quarter as before the hour.

Extra Practice, Unit 17, Exercise 2, pp. 183-184

Tests

Tests, Unit 17, 2A and 2B, pp. 115-118

Textbook

Page 73

Tasks 1-2, pp. 75-75

1. (a) 8
   (b) 3
   (c) 3
   (d) half past 6
   (e) 12
2. (a) cooking
   (b) eating
   (c) bedtime story

Workbook   123-124
Exercise 4, pp. ~~45-47~~

1. ● → close to 6 o'clock
   ● → a little after 11
   ● → almost 3 o'clock
   ● → about 8 o'clock
   ● → About half past 2
   ● → a little before 4
2. Watching a movie
3. Paining a picture

Review

Review

Workbook

Review 12, pp. 125-129

If your student has difficulties with the word problems in this review, get her to act them out, or draw number bonds.

Tests

Tests, Unit 17 Cumulative Tests A and B, pp. 119-126

Workbook

Review 12, pp. 125-129

1. (a) 28   (b) 40
2. (a) 6   (b) 30   (c) 35
3. 15; 25, 30, 35
4. (a) 15
   (b) 20
   (c) 20
   (d) 28
5. $30 + 9 = 39 = 40 - 1$
   $18 + 2 = 20 = 27 - 7$
   $20 + 8 = 28 = 38 - 10$
   $11 + 20 = 31 = 39 - 8$
6. Half past three   Half past four
    Three o'clock   Four o'clock
7. $6 + 4 + 5 = 15$; 15
8. $12 + 3 = 15$; 15
9. $14 - 10 = 4$; 4
10. $6 + 9 = 15$; 15
11. $20 - 8 = 12$; 12
12. • → 6:00
    • → 8:00
    • → 10:30

Unit 18 – Numbers to 100

Chapter 1 – Tens and Ones

Objectives

♦ Count by tens.
♦ Recognize number words for tens.
♦ Count within 100 by making tens.
♦ Read and write number words within 100.
♦ Interpret 2-digit numbers as tens and ones in terms of a part-whole model.

Material

♦ Counters
♦ Dot cards
♦ Numeral cards
♦ Number word cards for tens
♦ Base-10 blocks

Notes

In this chapter your student will deal with numbers up to one hundred. First he will count by tens, read and write numerals and number words for numbers to 100, and interpret 2-digit numbers within 100 in terms of tens and ones.

Your student should be able to both count objects within 100 as tens and ones and write the corresponding numeral, and "break apart" a 2-digit numeral into tens and ones.

Give your student as much practice as needed counting and making tens with units, such as straws, toothpicks, craft sticks, or multilink cubes, before using coins or base-10 blocks.

This section is an extension of what your student has learned in Unit 3 (Numbers to 40) so she should not have much difficulty with the concepts.

(1) Count by tens, recognize number words for tens

Textbook

Pages 76-78

   7; 8; 9; 10

Task 1, p. 79

  1. (a) 5 tens = 50
       (b) 6 tens = 60
       (c) 9 tens = 90

Workbook

Exercise 1, pp. 130-132

1. 80 = 8 tens = eighty
    40 = 4 tens = forty
    70 = 7 tens = seventy
    50 = 5 tens = fifty
2. 5; 50
    6; 60
    8; 80
    10; 100
3. 60   20
    90   80
    10   30
    50   100
    70   40

Discussion

Pages 76-77

Use base-10 material. Set out 4 tens and ask your student to say and write the number. Add another ten and get her to say and write the total amount if she can, otherwise tell her the next ten is fifty. Remind her that the 0 in the ones place means there are no ones, and the 5 is in the tens place so it means 5 tens. Continue to 100. You can tell her that 100 has a 1 in the hundreds place and that is the same as ten tens but you do not need to dwell on the hundreds place at this time.

Discussion

Pages 76-77

Ask your student how many beans are in each bundle on p. 76. Then ask her how many bundles there are on the bottom of p. 76. (4 bundles). Get her to count them as 1 ten, 2 tens, 3 tens, etc. and then as ten, twenty, thirty, forty. Write the numerals, 10, 20, 30, 40.

Have her count the bundles on pages ~~67-68~~ 76-78 in tens and read the numerals and number words.

Task 1, p. ~~75~~ 79

Workbook

Exercise 1, pp. 130-132

Reinforcement

Use ten dot-cards, each with 10 dots on them, or other base-10 type of material. Write a number word for tens, such as seventy, and ask her to show the correct number of dot cards and to write the corresponding numeral.

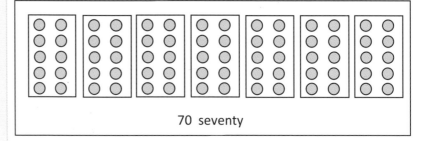

70 seventy

Mix up the numeral cards for tens (10, 20, 30 , 40, 50, 60, 70, 80, 90, and 100) and ask your student to put them in order. Do the same with the number words.

(2) Count within 100

Activity

Give your student a handful of counters (between 40 and 100) and discuss ways to count them. By now, he should think right away of putting them into groups of tens first. Have him count the groups of ten, and then count up from tens by the ones to get the total number.

Show the number with number cards, number bond, and on a place-value chart.

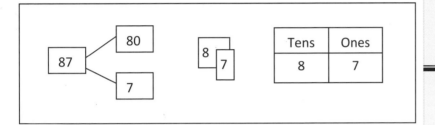

Repeat as needed.

Discussion

Tasks 2-3, pp. 80-81

Workbook

Exercises 2-3, pp. 133-136

Reinforcement

Use base-10 blocks. Write a number between 40 and 100 and get your student to show you the number using the base-10 blocks.

Textbook

Tasks 2-3, pp. 80-81

2. (a) 7 tens and 4 ones
   (b) 6 tens and 3 ones
3. (a) 5 tens and 3 ones = 53
   (b) 6 tens and 2 ones = 62
   (c) 4 tens and 2 ones = 42

Workbook

Exercise 2, pp. 133-134

1. 87   92
   34   64
   44   80
   75   76
2. Check answers

Exercise 3, pp. 135-136

1. 4 tens 4 ones; 44
   5 tens 2 ones; 52
   6 tens 3 ones; 63
   8 tens 5 ones; 85
2. (a) 46
   (b) 68
   (c) 83

(3) Read and write number words for numbers within 100

Activity

Make sure your student can read the following words:

one	eleven	ten	hundred
two	twelve	twenty	
three	thirteen	thirty	
four	fourteen	forty	
five	fifteen	fifty	
six	sixteen	sixty	
seven	seventeen	seventy	
eight	eighteen	eighty	
nine	nineteen	ninety	

Write a number between 40 and 100, and then write the number word. Remind your student that for tens and ones greater than 20, we write a little line (dash) between the word for the tens and the ones.

> 87
> Eighty-seven

Write some number words and get your student to read them and write the number.

Workbook

Exercise 4, pp. 137-138

Reinforcement

Mix up some number word cards, display them one at a time, and ask your student to read them and write the numeral.

Mix up some number word cards and ask your student to arrange them in order.

Workbook

Exercise 4, pp. 137-138

1. 63   59
   44   42
   55   68
   71   90
   100   87
2. 72   85
   93   74
   51   39
   28   12
   82   47

(4) Add ones to tens

Discussion

Task 4, p 82

This task illustrates that we add ones to tens to get a number with tens and ones.

Activity

Write some addition equations involving the addition of tens and ones and get your student to solve them. Use base-10 blocks to illustrate, if necessary, and/or use number bonds. For example:

⇒ 80 + 2 = _____

⇒ 60 + 4 = _____

⇒ 40 + _____ = 47

⇒ 50 + _____ = 51

⇒ _____ + 8 = 98

⇒ _____ + 5 = 55

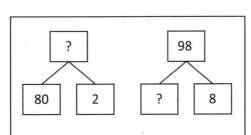

Workbook

Exercise 5, pp. 139-140

Reinforcement

Extra Practice, Unit 18, Exercise 1, pp. 175-178

Tests

Tests, Unit 18, 1A and 1B, pp. 127-130

Textbook

Task 4, p. 82

1. 71
2. 70 + 6 = 76
3. 80 + 8 = 88

Workbook

Exercise 5, pp. 139-140

1. 45
   57
   64
   72
2. 53   46
   66   57
   62   84

Chapter 2 – Estimation

Objectives

♦ Estimate quantities by comparing to a known quantity.
♦ Estimate quantity by making a reasonable guess.

Material

♦ Counters
♦ Jars or glasses (clear-sided)
♦ Other small items that can be held in hands or placed in bowls, e.g. pennies, cherries, marbles

Notes

In this chapter your student will estimate the quantity of an item by comparing the amount with a known amount.

Being able to make reasonable estimates based on the size of an object and experience indicates that a student has a "feel" for how large a number is. For example, if your student knows that 5 people can fit in a car, it makes sense that maybe 30 could fit in a bus, but not likely 100. Or, if she can hold about 10 cherries in her hand, then the bowl that is a bit larger might hold about 40, but not hundreds.

(1) Estimate quantities of objects

Activity

Use several jars or glasses of the same size. Put 10 counters in one and about 50 in another. Tell your student that there are 10 in the first glass and ask him to estimate, or guess, how many are in the other glass. He can then count to verify.

Ask your student how many counters he thinks he can hold in his hands, then let him test his estimate. Repeat with something larger, such as the multilink cubes.

Put a handful of counters on the table (less than 50). Ask your student to estimate how many there are, and then count to see how close she came. Repeat with a different amount.

Discussion

Tasks 1-3, pp. 83-84

Workbook

Exercise 6, p. 141

Reinforcement

Use opportunities that present themselves to estimate amounts within 100, such as about how many people are on the bus, or waiting in line at the store, or how many apples in a bag, or trees in the park.

Extra Practice, Unit 18, Exercise 2, pp. 179-180

Tests

Tests, Unit 18, 2A and 2B, pp. 131-135

Textbook

Task 1-3, pp. 83-84

1. Accept all reasonable answers.
   (a) 20
   (b) 40
   (c) 70
2. Accept all reasonable answers.
3. (a) 20
   (b) 19

Workbook

Exercise 6, p. 141

1. 22
2. 20
   24

Chapter 3 – Order of Numbers

Activity

♦ Understand order of numbers.
♦ Compare numbers within 100 using a number chart.
♦ Find the number that is 1 or 10 less or the number that is 1 or 10 more than a given number within 100.
♦ Count on and count back by tens and ones.

Material

♦ Hundred-chart
♦ Counters
♦ Number cubes
♦ Multilink cubes
♦ Hundred-chart (see appendix p. a12)

Notes

Students have already learned to count on by ones and count back by ones. Here your student will review this and also learn to count on by tens and count back by tens by relating a given number on the 100 chart with neighboring numbers.

Counting on by ones is useful for mentally adding 1, 2, or 3 to a number. Your student should be able to keep track of the three numbers without manipulatives or fingers, and this can be a quick strategy. He will use other strategies to add larger numbers.

Counting back by ones can be used to subtract ones from a number.

Counting on by tens is helpful for adding 10, 20, or 30 mentally to a number.

Counting back by tens is useful for mentally subtracting 10, 20, and 30 from a number.

The first two lessons in this section are primarily a review of concepts learned with numbers to 40, but extended to numbers to 100. Your student may not need the lessons, and can just do the workbook exercises.

$64 + 3$
Count on 3 ones from 64
65, 66, 67
$64 + 3 = 67$

$91 - 3$
Count back 3 ones from 91
90, 89, 88
$91 - 3 = 88$

$45 + 30$
Count on 3 tens from 45
55, 65, 75
$45 + 30 = 75$

$77 - 30$
Count back 3 tens from 77
67, 57, 47
$77 - 30 = 47$

(1) Understand order of numbers

Activity

Use a hundred-chart. Cover up some of the numbers with opaque counters. Ask your student what number comes after the previous number or before the succeeding number. For example, if 47 is covered up, ask your student what number comes after 46 or before 48.

Point to two different numbers on the chart that are in the same row. Ask your student which is greater and then ask her to explain why she chose that one. It has more ones. Do the same for two numbers in the same column, and then for two numbers in different columns and rows.

Write 4 numbers within 100 on some index cards and ask your student to put them in order. Remind him, if needed, that he needs to first put them in order according to the tens digit, and then, if two or more have the same tens, put them in order according to the ones digit. If necessary, he can look at their position on the hundreds chart.

Workbook

Exercises 7-8, pp. 142-143

Allow your student to use a hundred-chart when doing Exercise 8.

Reinforcement

Dot-to-dot pictures are a good way to reinforce number order. Most children are more interested in finding the next number if they know there will be a picture at the end. There are sites on the internet that have interactive dot-to-dots and printable pages – you can find them by putting "dot-to-dot" in a search engine on the internet.

Workbook

Exercise 7, p. 142

1. (a) 67, 68; 70;
       72, 73; 75, 76;
       79, 80; 82
   (b) 93, 92; 90, 89; 87;
       85; 81, 80

Exercise 8, p. 143

1. 17, 36, 43; 12; 43
   50, 38, 29; 29; 52
   86, 84, 68; 68; 93
   60, 72, 95; 58; 95

(2) Count on or count back one or ten

Activity

Use base-10 material, such as multilink cubes in tens and ones. Display a number within 90, such as 56, and ask your student for the number that is 1 more and 10 more, then 1 less, and 10 less than the number displayed. Write the equations.

Write some equations involving the addition or subtraction of 1 or

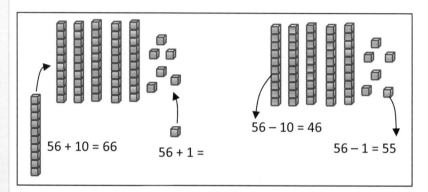

$$56 + 10 = 66$$
$$56 + 1 =$$
$$56 - 10 = 46$$
$$56 - 1 = 55$$

10, and get your student to solve them. Make sure she can determine when to change the tens, and when to change the ones. We add 10 to the tens of the 1-digit number, not the ones.

$\Rightarrow \quad 66 + 1 = $ ___

$\Rightarrow \quad 55 - 1 = $ _____

$\Rightarrow \quad 77 + 10 = $ ____

$\Rightarrow \quad 88 - 10 = $ ____

$\Rightarrow \quad 48 + $ ___ $ = 49$

$\Rightarrow \quad 53 - $ ___ $ = 52$

$\Rightarrow \quad 87 - $ ___ $ = 77$

$\Rightarrow \quad $ ___ $ + 10 = 63$

Workbook

Exercise 9-10, pp. 144-146

Workbook

Exercise 9, pp. 144-145

1. 15; 24
   23; 32   44; 53
   36; 45   57; 66
2. 22; 13
   14; 5   38; 29
   26; 17   50; 41

Exercise 10, p. 146

1. 59   51 → 52
    ↓   ↑   ↓
    69   41   62
    ↓   ↑   ↓
    68   40   61
    ↓   ↑   ↓
    58   50   71
    ↓   ↑   ↓
    57   49   70
    ↓   ↑   ↓
    47 → 48 80

(3) Count on and count back by tens and ones

Activity

Pages 85-87

Use a hundred-chart. You may want to use different color counters to represent the different animals and put them on a hundreds chart in order to let your student move the counters as he counts back.

After doing the activity on this page, continue to have your student to practice with the hundred-chart. Point to a number and ask your student to count forward or backward a specified number of tens or ones. For example, put the counter on 55 and ask your student to count on 4 tens. Include situations where the counter has to move to the start of the next row or the end of the previous row when counting on or back by ones. For example, put a counter on 42 and ask the student to count back 5 ones.

Discussion

Tasks 1-3, p. 88

Workbook

Exercises 11-12, pp. 147-149

Reinforcement

Say a number between 0 and 10 and have your student count on from that number by tens to the 90s. For example, 4, 14, 24, 34, 44, 54, 64, 74, 84, 94. Say a number between 90 and 100 and ask your student to count back by tens from that number to the ones, e.g. 97, 87, 77, 67, 57, 47, 37, 27, 17, 7.

Circle two numbers and ask your student to count from one to the next by tens and then by ones, both forward and backward, with help from the hundred-chart. For example, the two numbers are 37 and 82. Your student counts on, 47, 57, 67, 77, 78, 79, 80, 81, 82 or back, 72, 62, 52, 42, 41, 40, 39, 38, 37.

Write some addition or subtraction equations involving adding or subtracting 1, 2, 3, 10, 20, or 30 and ask your student to solve them. See if he can do it without the hundred-chart.

⇒ $48 + 3 =$ ___
⇒ $54 - 2 =$ ___
⇒ $35 + 30 =$ ____
⇒ $87 - 20 =$ ____
⇒ $62 +$ ___ $= 65$
⇒ $74 -$ ___ $= 54$

Textbook

Pages 85-87

Tasks 1-3, p. 88

1. (a) 65
   (b) 63
   (c) 74
   (d) 54
2. (a) 50, 60, 80
   (b) 56, 66, 76
   (c) 53, 33, 23
3. (a) 62
   (b) 72
   (c) 86
   (d) 76

Workbook

Exercise 11, p. 147-148

1. (a) 55
   (b) 69
   (c) 75
   (d) 66
   (e) 74
   (f) 26
   (g) 84
   (h) 84

Exercise 12, p. 149

1. (a) 78   (b) 87
   (c) 81   (d) 72
   (e) 81   (f) 78
   (g) 81   (h) 76
   (i) 78   (j) 92
   (k) 76

Reinforcement

Label 2 cubes, one with +, +, +, −, −, − and the other with 1, 2, 3, 10, 20, 30. Use a counter and a hundred-chart. Your student puts the counter on 55. She then throws both cubes and moves as indicated by the throw. For example, if she throws + and 30, she moves 3 tens forwards. Let her see how long it takes to end up off the board (for example, if she is on 14 and throws − and 20). You can also play this as a game, to see who gets off the board first. You can ask her to write a number equation for each move.

Extra Practice, Unit 18, Exercise 3, pp. 181-182

Tests

Tests, Unit 18, 3A and 3B, pp. 137-139

Chapter 4 – Comparing Numbers

Objectives

Material

♦ Base-10 blocks or multilink cubes in tens and ones
♦ Playing cards or 4 sets of number cards 0-9

Notes

In the previous lessons, students learned to compare and order numbers based on their position in a number chart. In this chapter, they will compare numbers by focusing on the tens and ones digits of the numbers they are comparing. The symbols for greater than, ">", and less than,"<", are introduced.

Numbers can be compared by comparing the digits in the highest place value first. If they are the same, then we compare the digits in the next highest place value, and so on. This process will be extended in later levels to numbers past 100. For numbers within 100, we first compare the tens. If they are the same, then we compare the ones. Students will do that first with base-10 material, and then by simply looking at the numbers. They need to realize that they are comparing tens, not just the first digit in the number. With numbers that have only 1 or two digits, this is fairly obvious from being familiar with a 100-chart and the order of numbers on it, but with numbers with more digits, such as students will encounter in later grades, it may be less obvious.

Compare 21 and 12.	Compare 32 and 35.	Compare 25 and 8.
2 1	3 2	2 5
1 2	3 5	8
2 tens > 1 ten,	3 tens = 3 tens	2 tens > 0 tens
so 21 > 12	2 ones < 5 ones	So 25 > 8
	So 32 < 35	

(1) Use the symbols for greater than and less than

Textbook

Page 89

Tasks 1-3, p. 90

1. (a) 43 > 34
   (b) 69 < 78
   (c) 35 > 32
   (d) 29 < 37
   (e) 50 > 49
2. (a) 39   (b) 30
   (c) 56   (d) 98
3. 50, 59, 90, 95

Workbook

Exercise 13, pp. 150-152

1. (a) 50   (b) 59
   (c) 28   (d) 70
   (e) 87   (f) 100
2. (a) 45   (b) 87
   (c) 63   (d) 100
   (e) 70   (f) 57
3. (a) 25   (b) 30
   (c) 31   (d) 87
   (e) 57   (f) 89
   (g) 63   (h) 100
4. (a) 35   (b) 59
   (c) 50   (d) 66
   (e) 26   (f) 40
5. (a) 67, 76, 78, 87
   (b) 90, 82, 79, 66
6. (a) >   (b) <
   (c) >   (d) <
   (e) <   (f) >
   (g) >   (h) <
   (i) <   (j) >

Activity

Ask your student what sign, or *symbol*, we use to show that two numbers or two expressions are the same. (Equal sign.) Tell your student that we can also use symbols to show that one number is greater than or less than another.

Discussion

Page 89

The symbols ">" and "<" can be remembered by associating them with the mouth of a greedy crocodile that eats the greater number. Point out that 21 is greater than 12 because it has more tens. 99 is less that 100 because 99 has 9 tens but 100 has 10 tens.

Activity

Write two 3-digit numbers with different tens, one above the other, with the digits aligned. Have your student show both numbers with base-10 material. (Do not use dimes and nickels.) Point out that the number with more tens is larger. Write the numbers next to each other with the symbols.

Repeat with two number with the same tens digit but different ones digit. Tell your student that we first compare the tens. If the tens are the same, we then compare the ones to see which number is greater.

Repeat with a 1-digit number and a 2-digit number. Point out that the first digit of the 1-digit number is greater, but it is ones. The 9 has no tens. 12 is greater because it has more tens.

Write some numbers next to each other and have your student write the symbols for "greater than", "less than", or "equal to" between them.

Discussion

Tasks 1-3, p. 90

42
62
42 < 62
62 > 42
56
53
56 > 53
53 < 56
9
12
9 < 12
12 > 9
32 < 59
60 > 38
46 < 48
82 = 82
9 < 17
100 > 10

Reinforcement

Write three or more 2-digit numbers on pieces of paper or index cards and get your student to arrange them in order.

If your student has problems remembering which symbol to use, draw the symbols on half-pieces of index cards with crocodiles and open mouths. Also draw the plain symbols on two half-index cards. Write two numbers with space between them, or write them on index cards and put them next to each other with a space between them. Have your student put the correct symbol between them, first using the crocodile cards and then using the cards with plain symbols.

Game

Material: A deck of cards with face 10 and face cards removed, or four sets of number cards 0-9
Procedure: Each player draws two cards. The first card is the tens, the second the ones. They then compare their numbers. The player with the greater number gets a point, or gets both cards. The player with the most points at the end, or more cards, wins.

Extra Practice, Unit 18, Exercise 4, pp. 183-184

Tests

Tests, Unit 18, 4A and 4B, pp. 141-144

Chapter 5 – Addition Within 100

Objectives

♦ Add 2-digit numbers. Mentally.
♦ Understand the vertical representation for addition problems.

Material

♦ Base-10 blocks
♦ Hundred-chart
♦ Multilink cubes
♦ Number cubes
♦ Playing cards
♦ Mental Math 18-22

Notes

In unit 3 (Numbers to 40) students learned to add a 1-digit number to a 2-digit number with or without renaming. This is reviewed here and extended to numbers within 100.

Students should recognize when adding a 1-digit number to another number will increase the tens, that is, whether the sum of the ones is greater than ten. They can add either by recalling the addition fact, or by using the "make 10" strategy.

$$67 + 8 = 60 + 15 = 75$$
$$/\ \backslash$$
$$60 \quad 7$$

$$67 + 8 = 70 + 5 = 75$$
$$/\ \backslash$$
$$3 \quad 5$$

When the 1-digit number is 1, 2, or 3, they may add by counting on by ones.

$$68 + 3 = 71$$
69, 70, 71

When the 2-digit number is tens, they can simply write the answer using place-value concept.

$$30 + 60 = 90$$
$$(3 + 6 = 9, 3 \text{ tens} + 6 \text{ tens} = 9 \text{ tens})$$

In this section, students will learn a new skill, adding a 2-digit number to a 2-digit number.

First they will learn to add tens.

$$32 + 60 = 92$$
$$/\ \backslash$$
$$2 \quad 30$$

Then they will add a 2-digit number by simply adding the tens first, then adding the ones.

$$32 + 64 = 64 + 32 = 96$$
$$/\ \backslash$$
$$30 \quad 2$$

You can draw arrow diagrams to illustrate the process. Remind your student that addition can be done in any order, and encourage her to add the smaller number to the larger.

$$64 \xrightarrow{+30} 94 \xrightarrow{+2} 96$$

Your student will learn the vertical representation for addition expressions in which the numbers are written one below the other, aligning the digits with the same place value. None of the problems with the vertical form will involve renaming, so he can add tens first or ones first. Do not insist that he add ones first when seeing a vertical format at this level. The formal algorithm for adding numbers, in which the numbers are written in vertical form, and ones are added first, renaming if necessary, will be taught in *Primary Mathematics* 2A.

Your student should by now know the addition facts through 10, and most of the facts through 20. Some students need additional practice. Note that some students who are very good at math concepts have difficulty memorizing math facts. Math comprehension and memorization skills are not related. However, speed and accuracy with math facts is useful. If your student has trouble memorizing math facts, don't over-emphasize instant recall and let your student calculate them using various strategies, such as making a ten. Knowing these strategies will give her a tool to help find the fact fast if he or she forgets it or is not sure of the answer.

As you do these units, be sure your student is proficient at recalling or calculating the following math facts.

9 + 2	9 + 3	9 + 4	9 + 5	9 + 6	9 + 7	9 + 8	9 + 9
	8 + 3	8 + 4	8 + 5	8 + 6	8 + 7	8 + 8	8 + 9
		7 + 4	7 + 5	7 + 6	7 + 7	7 + 8	7 + 9
			6 + 5	6 + 6	6 + 7	6 + 8	6 + 9
				5 + 6	5 + 7	5 + 8	5 + 9
					4 + 7	4 + 8	4 + 9
						3 + 8	3 + 9
							2 + 9

Use drill sheets, fact cards, computer games, or games such as the following:

Material: Playing cards, with face cards removed. The Ace is 1.
Procedure: Play a game similar to "War". Shuffle and deal all cards out. Each player turns over two cards and adds the numbers on their card. The player with the highest total gets all the cards. After all cards have been turned over, the player with the most cards wins.

Material: Two number cubes labeled 5-10, counters, a game-board with the numbers 10-20 written randomly on a grid. (You can use as large a grid as you like. For example, use a wet-erase marker and write the numbers 10-20 randomly in the blank squares on the back of a laminated hundreds board.)
Procedure: Players take turns throwing the number cube, adding the numbers, and putting their counter on the answer on the game board. The first player to get three counters in a line (vertically, horizontally, or diagonally) wins.

(1) Add ones without renaming

Textbook

Page 91

Task 1, p. 92

1. 65 + 2 = 67

Activity

Write the expression 62 + 3. Ask your student to solve it. Discuss two methods:

⇒ Count on by ones from 62: 63, 64, 65
⇒ Add the ones

Illustrate adding the ones with number bonds and with base-10 material if necessary. Show how 62 can be split into 60 and 2, and the 2 added to 3 to give 5 ones, so that the total is now 60 and 5, or 65. You can show the number bonds similar to how they are shown in the textbook (see p. 77) or sideways, showing 3 being added to the 2.

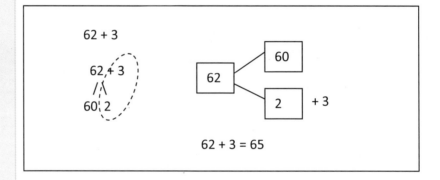

Write the expression: 74 + 5. Ask your student to solve it. Point out that when adding ones greater than 3, it is harder to keep track of how many ones have been added (without fingers) and it takes longer than simply remembering 4 + 5 = 9 and adding ones first.

74 + 5 = 79
/\
70 4

Repeat with other examples as needed.

Discussion

Page 91

Task 1, p. 92

If your student needs additional review, refer back to the lessons under part 3 of unit 3. You can use the same teaching activities for addition, but use numbers past 40.

Workbook

Exercise 14, pp. 153-154

Workbook

Exercise 14, pp. 153-154

1. (a) 27
   (b) 37
   (c) 19
   (d) 48
2. 37
   7; 27   7; 37
   8; 48   9; 59
   6; 66   9; 79

(2) Add ones with renaming

Activity

Write the expression 78 + 5. Show the two numbers with base-10 material.

Ask your student if adding the ones will result in ten or more. Discuss two methods for finding the solution. Use base-10 material and number bonds to illustrate these methods.

- Add the ones first

8 + 5 = 13
70 + 13 = 83

- Make tens and add the remaining ones

78 + 2 = 80
80 + 3 = 83

78 + 5

78 + 5
/\
70 8

78 + 5 = 83

78 + 5
/\
2 3

78 + 5 = 83

Discussion

Task 2, p. 92

Repeat with other examples as needed.

Workbook

Exercise 15, pp. 155-156

Textbook

Task 2, p. 92

2. (a) 74 + 6 = 80

Workbook

Exercise 15, pp. 155-156

1. 50
   80
   63
   91
2. 72
   10; 50   11; 71
   14; 94   13; 63
   11; 81   10; 70

Reinforcement

Use a deck of cards. Shuffle. Turn the two cards over; the first is the tens and the second is the ones of a 2-digit number. Turn a third card over. This is the ones to be added. Ask your student to add the two numbers. If the total is greater than 100 (e.g. 9, 8, and 9 are turned over, 98 + 9 = 107) replace the first card to the bottom of a pile and draw again. (The second card is now the first card, and the new card is the third card.) This can be played as a game by dealing all cards out, and having each player turn over three cards to make the two numbers. The player with the greatest total wins the round.

Mental Math 18-19

(3) Add tens

Activity

Write the expression 50 + 20. Show the two numbers with base-10 material.

Ask your student to solve it. Point out that she can count on by tens (60, 70) or remember that 5 + 2 = 7, so 5 tens + 2 tens = 7 tens.

Add some ones to the first number: 53 + 20. Point out that adding the ones does not change the total tens. We can find the answer to 53 + 20 by simply adding 50 and 20, and writing down the same number of ones.

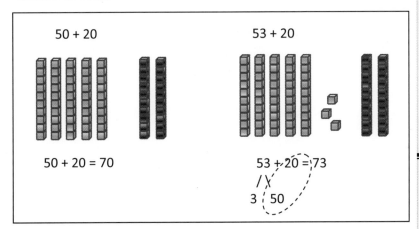

Write the expression: 20 + 53. Point out that since we can add in any order, we can solve this in the same way as 53 + 20.

$\Rightarrow$ 53 + 20 = 73
$\Rightarrow$ 20 + 53 = 73

Discussion

Tasks 3-4, pp. 93-94

Provide additional examples if needed.

Workbook

Exercise 16-17, pp. 157-160

Reinforcement

Mental Math 20

Textbook

Tasks 3-4, pp. 93-94

3. 62 + 30 = 92
4. (a) 43 + 20 = 63
    (b) 74 + 10 = 84
    (c) 30 + 46 = 76

Workbook

Exercise 16, pp. 157-159

1. 50
    50
    60
    80
2. 8; 80
    5; 50   6; 60
    7; 70   8; 80
    9; 90   6; 60
    9; 90   9; 90

Exercise 17, pp. 159-160

1. 64
    76
    89
    97
2. 62
    70; 76   90; 98
    70; 77   80; 85
    70; 74   80; 81

(4) Add 2-digit numbers

Textbook

Tasks 5-6, p. 95

5. 32 + 16 = 48
6. 43 + 35 = 78

Activity

Write the expression 45 + 32. Show the numbers with base-10 material. Show your student that we can first add the tens (by moving the tens over) and then the ones (by moving the ones over).

Draw an arrow diagram as you ask your student for the answer to 45 + 30, and then 45 + 32.

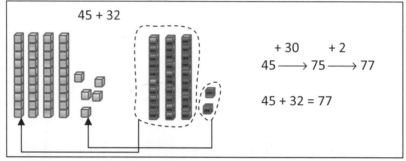

Repeat with 28 + 61. Point out that it is often easier to add the smaller number to the larger number, so we can solve this as 61 + 20 + 8.

Write the expression 38 + 26. Show the numbers with base-10 material. First the tens are added; 38 + 20 = 58. Then the ones are added: 58 + 6 = 64. You can show the steps with an arrow diagram.

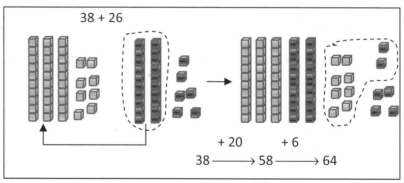

Provide additional practice where the ones add to ten or more.

Discussion

Tasks 5-6, p. 95

Workbook

Exercise 18, 161-162

Reinforcement

Mental Math 21-22

Workbook

Exercise 18, 161-162

1. 39
    58; 58   36; 36
    50; 50   90; 90
    42; 42   64; 64
2. 58
    78; 78   77; 77
    80; 80   90; 90
    75; 75   94; 94

(5) Understand the vertical representation for addition problems

Activity

Draw a place-value chart and write 25 above 3 on the chart, aligning the digits. Do the same for 25 and 10. Point out that the tens and ones of each number are lined up, one below the other. Add a "+" before the second row. Sell your student that this shows that we are adding the two numbers together. Draw a line below the second row and tell her that instead of an equal sign we draw a line and then put the answer below the line. Show that ones are added to ones, and tens are added to tens.

Tens	Ones
2	5
+	3
2	8

Tens	Ones
2	5
+ 1	0
3	5

Discussion

Tasks 7-12, pp. 96-99

In tasks 7 and 10, ones are added first and then tens. Tell your student that when we write the problem vertically like this, it is easy to add ones first if we want to. We do not have to write tens and ones above the numbers, we just have to line them up correctly when we write the problem like this.

Write a few addition problems horizontally and get your student to practice writing them vertically, aligning the digits. Use problems where the sum of the ones is less than ten. If she has trouble aligning the digits, use graph paper or lined notebook paper turned sideways. There is some graph paper in the appendix that can be copied

In the workbook exercise, use your discretion for how many problem you want to make your student rewrite vertically, if any. The purpose of this lesson is simply to recognize what to do with problems written vertically. The emphasis should be on mental math at this point.

Workbook

Exercise 19, 163-166

Reinforcement

Extra Practice, Unit 18, Exercise 5, pp. 185-188

Tests

Tests, Unit 18, 5A and 5B, pp. 145-148

Textbook

Tasks 7-12, pp. 96-99

8. 58
9. 75
11. 58
12. 89

Workbook

Exercise 19, 163-166

1. (a) 67   (b) 35
   (c) 48   (d) 79
2. (a) 34   (b) 59
   (c) 77   (d) 55
3. (a) 66   (b) 54
   (c) 49   (d) 46
   (e) 83   (f) 99
4. (a) 89   (b) 97
   (c) 58   (d) 99
   (e) 95   (f) 95

Chapter 4 – Subtraction Within 100

Objectives

♦ Subtract 2-digit numbers mentally.
♦ Understand the vertical representation for subtraction problems.

Material

♦ Base-10 blocks
♦ Multilink cubes
♦ Playing cards
♦ Mental Math 23-27

Notes

In unit 3 (Numbers to 40) students learned to subtract a 1-digit number from a 2-digit number with or without renaming. This is reviewed here and extended to numbers within 100.

Your student should recognize whether subtracting the 1-digit number from another number will decrease the tens, that is, there are not enough ones. If there are enough ones, they simply subtract the ones.

$$68 - 6 = 62$$
$$/\ \backslash$$
$$60 \quad 8$$

When the 2-digit number is tens, they can use the "subtract from ten" strategy.

$$60 - 8 = 52$$
$$/\ \backslash$$
$$50 \quad 10$$

If there are not enough ones, they can subtract either by recalling the subtraction fact, or by using the "subtract from 10" strategy.

$$62 - 8 = 50 + 4 = 54$$
$$/\ \backslash$$
$$50 \quad 12$$

$$62 - 8 = 2 + 52 = 54$$
$$/\ \backslash$$
$$2 \quad 60$$

When the 1-digit number is 1, 2, or 3, they may subtract by counting back by ones.

$$52 - 3 \qquad 51, 50, 49$$
$$52 - 3 = 49$$

In this section, students will learn a new skill, subtracting a 2-digit number from a 2-digit number. First they will learn to subtract tens.

$$90 - 30 = 60$$
$$(9 - 3 = 6, 9 \text{ tens} - 3 \text{ tens} = 6 \text{ tens})$$

$$98 - 30 = 68$$
$$/\ \backslash$$
$$8 \quad 90$$

Then they will subtract a 2-digit number by simply subtracting the tens first, and then subtracting the ones. You can draw arrow diagrams to illustrate the process.

$$98 - 32 = 68 - 2 = 66$$
$$/\ \backslash$$
$$30 \quad 2$$

$$98 \xrightarrow{-30} 68 \xrightarrow{-2} 66$$

Students will see the vertical representation for

subtraction problems in this chapter. The problems shown this way will not involve renaming. The formal algorithm for subtracting numbers, in which the numbers are written in vertical form, and ones are subtracted first, renaming tens if necessary, will be taught in *Primary Mathematics* 2A. Do not require your student to add ones first at this level.

Your student should by now know the subtraction facts through 10, and either know the facts through 20 or be able to compute them quickly using subtraction from a ten. As you do these units, be sure your student is proficient at recalling or calculating the following math facts.

11 – 10	11 – 9	11 – 8	11 – 7	11 – 6	11 – 5	11 – 4	11 – 3	11 – 2
12 – 10	12 – 9	12 – 8	12 – 7	12 – 6	12 – 5	12 – 4	12 – 3	
13 – 10	13 – 9	13 – 8	13 – 7	13 – 6	13 – 5	13 – 4		
14 – 10	14 – 9	14 – 8	14 – 7	14 – 6	14 – 5			
15 – 10	15 – 9	15 – 8	15 – 7	15 – 6				
16 – 10	16 – 9	16 – 8	16 – 7					
17 – 10	17 – 9	17 – 8						
18 – 10	18 – 9							
19 – 10								

(1) Subtract ones without renaming

Textbook

Page 100

$48 - 2 = 46$

Task 1, p. 101

1. (a) $57 - 3 = 54$
   (b) $64 - 4 = 60$

Activity

Write expression 65 – 2. Show the 65 with base-10 material. Ask your student to solve it. Discuss two methods:

⇒ Count back by ones from 65: 64, 63.

⇒ Subtract ones.

Illustrate with number bonds and with base-10 material if necessary. Show how 65 can be split into 60 and 5, and the 2 subtracted from 5 ones, so that the total is now 60 and 3, or 63. You can show the number bonds similar to how they are shown in the textbook (see p. 77) or sideways, showing 3 being subtracted from the 5.

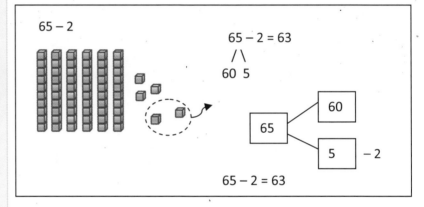

Write the expression 78 – 5 and ask your student to solve it. Point out that when subtracting ones greater than 3, it is harder to keep track of how many ones have been added (without fingers) and it takes longer than simply remembering 8 – 5 = 3 and subtracting the ones.

$$78 - 5 = 73$$
$$/ \backslash$$
$$70 \ 8$$

Repeat with other examples as needed.

Discussion

Page 100

Task 1, p. 101

If your student needs additional review, refer back to the lessons under part 3 of unit 3. You can use the same teaching activities for subtraction, but use numbers past 40.

Workbook

Exercise 20, pp. 167-168

Workbook

Exercise 20, pp. 167-168

1. 52
   77
   41
   64
2. 34
   2; 62
   3; 73
   4; 84
   3; 33
   2; 52
   4; 64

(2) Subtract ones with renaming

Activity

Write the expression: 70 – 5. Show 70 with base-10 material. Ask your student for the answer. She should recall that 10 – 5 = 5, realize that there is one less tens, and give the answer 65. If not, illustrate with base-10 material, moving one ten over and asking your student for 10 – 5. Then provide additional problems to be sure your student can easily subtract ones from tens.

Write the expression 73 – 5 and display 73 with base-10 material.

Ask your student if there are enough ones to take away 5 ones. There are not. Discuss two methods for finding the solution:

- Find the answer to 70 – 5. Replace one of the tens with 5 ones. There are now 65 and 3 ones still, so the answer is 68.

- Find the answer to 13 – 5. Replace 13 with 8 ones. There are still 6 tens, so the answer is 68.

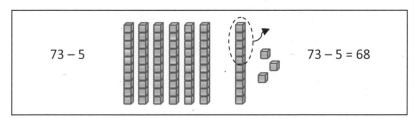

Write the expression 92 – 7. This time, while discussing the two methods, show the process with number bonds.

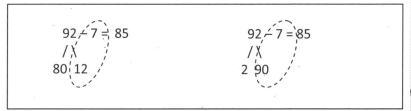

Repeat with other examples as needed.

Discussion

Task 2, textbook pp. 101-102

Get your student to explain his reasoning in the method he uses to subtract.

Workbook

Exercise 21, pp. 169-170

Textbook

Task 2, pp. 101-102

 2. (a) 60 – 3 = 57
    (b) 82 – 6 = 76
    (c) 53 – 6 = 47
    (d) 85 – 9 = 76

Workbook

Exercise 21, pp. 169-170

1. 78
   84
   37
   68
2. 38
   7; 47   8; 58
   6; 56   5; 65
   5; 75   7; 87

Reinforcement

Use a deck of cards with face cards and 10 removed, or 4 sets of number cards 1-9. Shuffle. Turn the two cards over; the first is the tens and the second is the ones of a 2-digit number. Turn a third card over. This is the ones to be subtracted. Ask your student to subtract the two numbers. This can be played as a game by dealing all cards out, and having each player turn over three cards to make the two numbers. The player with the greatest total wins the round.

Mental Math 23-27

(3) Subtract tens

Activity

Write the expression 80 – 30. Show 80 with base-10 material. Ask your student to find the answer. Point out that he can count back by tens (70, 60, 50) or remember that 8 – 3 = 5, so 8 tens – 3 tens = 5 tens.

Add some ones to the first number: 84 – 30. Illustrate with base-10 material. Point out that adding the ones does not change the total tens. We can find the answer to 84 – 30 by simply subtract the tens, and writing down the same number of ones.

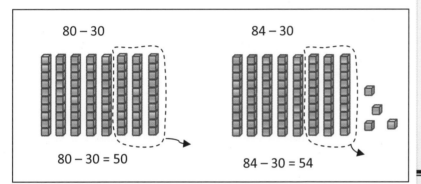

80 – 30	84 – 30
80 – 30 = 50	84 – 30 = 54

Write the expression 95 – 40. This time, show the process with number bonds by splitting 90 into 90 and 5.

Repeat with other examples as needed.

```
95 – 40 = 55
     / \
    5   90
```

Discussion

Tasks 3-4, pp. 103-104

Workbook

Exercises 22-23, pp. 171-174

Reinforcement

Mental Math 25

Textbook

Tasks 3-4, pp. 103-104

3. 53 – 20 = 33
4. (a) 73 – 10 = 63
    (b) 65 – 40 = 25
    (c) 94 – 30 = 64

Workbook

Exercise 22, pp. 171-172

1. 30
    20
    10
    50
2. 4 tens, 40

3; 30	2; 20
1; 10	3; 30
2; 20	5; 50
5; 50	3; 30

Exercise 23, pp. 173-174

1. 28
    27
    15
    31
2. 14

10; 19	40; 42
20; 26	40; 43
10; 17	10; 15

(4) Subtract 2-digit numbers

Textbook

Tasks 5-6, p. 105

  5. 56 – 14 = 42
  6. 78 – 32 = 46

Activity

Write the expression 67 – 32. Show the number 67 with base-10 material. Show your student that we can first subtract the tens (by removing 3 tens) and then the ones (by removing 2 ones).

Draw an arrow diagram as you ask your student for the answer to 67 – 30, and then 37 – 2.

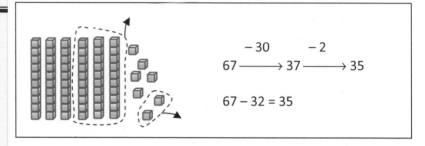

Write the expression 62 – 37 and illustrate with base-10 material.

Get your student to first subtract tens, 62 – 30 = 32, and then ones, 32 – 7 = 25. Draw the arrow diagram.

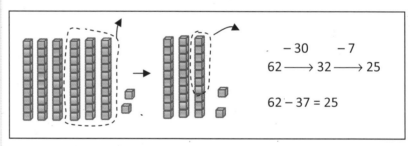

Provide additional examples as needed

Discussion

Tasks 5-6, p. 105

Workbook

Exercise 24, pp. 175-176

Reinforcement

Mental Math 26-27

Workbook

Exercise 24, pp. 175-176

  1. 23
      35; 35   52; 52
      40; 40   46; 46
      26; 26   48; 48
  2. 24
      44; 44   33; 33
      20; 20   32; 32
      26; 26   9; 9

(5) Understand the vertical representation

Activity

Write 27 – 5 vertically, aligning the digits. Do the same for 27 – 10. Show that we subtract ones from ones and tens from tens. Tell your student that sometimes subtraction problems are written this way. Point out the "–" sign which tells us that this is a subtraction problem and not an addition problem. When we write the problem this way, it is easy to subtract the ones first.

$$
\begin{array}{r} 2\ 7 \\ -\ \ \ 5 \\ \hline 2\ 2 \end{array}
\qquad
\begin{array}{r} 2\ 7 \\ -1\ 0 \\ \hline 1\ 7 \end{array}
$$

Discussion

Tasks 7-12, pp. 106-109

In tasks 7 and 10, ones are added first and then tens. Your student can solve the problems mentally and subtract tens first. Rewrite the problems in tasks 8-9 and 11-12 vertically, or have your student rewrite them, and get your student to write the answers aligning the digits in the correct places.

Write a few subtraction problems horizontally and get your student to practice writing them vertically, aligning the digits, and then finding the answer. Use problems where there is no renaming; that is, there are enough ones to subtract from.

Again, use your discretion in requiring your student to rewrite the problems in the exercise vertically. You could write them vertically for your student and let her write the answer. The skill in rewriting problems vertically can wait until next year; the purpose of this lesson is that your student understands the vertical representation rather than solving the problem in a particular way. At this level, the emphasis should still be on mental math for subtracting 2-digit numbers.

Workbook

Exercise 25, pp. 177–180

Reinforcement

Extra Practice, Unit 18, Exercise 6, pp. 189-192

Tests

Tests, Unit 18, 6A and 6B, pp. 149-152

Textbook

Tasks 7-12, pp. 106-109

8. 51
9. 64
11. 53
12. 53

Workbook

Exercise 25, pp. 177-180

1. (a) 43   (b) 51
   (c) 74   (d) 79
2. (a) 25   (b) 26
   (c) 44   (d) 33
3. (a) 26   (b) 41
   (c) 26   (d) 20
   (e) 42   (f) 42
4. (a) 11   (b) 44
   (c) 21   (d) 24
   (e) 45   (f) 13

Review

Review

Workbook

Review 13, pp. 181-185

Use the review in the workbook as an assessment to see if you need to re-teach any concepts.

For problem 13, you can have your student write the correct symbol (> or <) between the numbers.

Continue with mental math practice.

Tests

Tests, Unit 18 Cumulative Tests A and B, pp. 153-162

Workbook

Review 13, pp. 181-185

1. (a) 49
   (b) 58
2. (a) 60; 80; 100
   (b) 10, 12; 18, 20
   (c) 30, 25, 20
3. 69, 78, 84, 91, 100
4. 14; 14
5. 4
6. 4
7. $8 - 5 = 3$; 3
8. $12 - 2 = 10$; 10
9. $20 - 6 = 14$; 14
10. $8 + 6 = 14$; 14
11. $15 - 4 = 11$; 11
12. (a) 66 (<)
    (b) 98 (>)
    (c) 77 (>)
    (d) 65 (<)
13. (a) Set B
    Accept reasonable answers for (b) and (c)
    (b) 35
    (c) 10

Unit 19 – Money

Chapter 1 – Bills and Coins

Objectives

- Recognize and name coins and bills up to $20.
- Change a coin or bill for an equivalent set of coins or bills of a smaller denomination.
- Count the amount of money in a set of coins or a set of bills.
- Understand the symbols ¢ and $.
- Make up a set of coins or bills for a given amount.
- Compare the amount of money in two or three sets of coins or bills.
- Compare the price of two items in cents or dollars.
- Count by fives.

Material

- Coins
- Bills

Notes

Your student is probably already familiar with coins and bills. In this chapter, he will learn to recognize and name coins and bills up to a $10 bill. They will count and compare different amounts of money in coins or in bills. He will also make different combinations of coins to make up a coin of larger denomination or a dollar bill. Use actual money to help them recognize and name the coins and bills.

Your student will learn to recognize the symbols ¢ (cent) and $ (dollar). At this stage, she will only deal with sets of coins or sets of dollars. Combinations of dollars and cents and notations such as $3.40 will be introduced in *Primary Mathematics 2*.

In this chapter, your student will practice counting by 5's. Counting by 5's is necessary to count nickels.

When possible, show your student commemorative coins, such as state quarters or dollar coins. 50 cent coins are used here, so include them if possible, though they are no longer very common.

(1) Recognize coins and bills and their value

Textbook

Pages 110

Tasks 1-4, pp. 111-114

1. 10; 25
2. (a) 5
   (b) 2; 10
   (c) 5; 25
3. (a) 10
   (b) 10
   (c) 5
   (d) 10
4. (a) 20; 25
5. (a) 10
   (b) 35
   (c) 50
6. 100; 1

Workbook

Exercise 1, p. 186

1. 5, 10, 15. 20, 25, 30, 35,
   40, 45, 50, 55
   55
2. 100

Activity

Show your student various coins and bills and discuss their appearance and values. Tell your student that a dollar is the same as 100 cents.

Show your student a dime and ask how many pennies have the same value. Make sure your student understands that to pay for something that costs 10 cents, she can use either a dime or ten pennies.

Ask your student how many nickels make up a dime. Ask her if there is another way to use nickels and pennies to make up a dime (1 nickel, 5 pennies).

Set out 1 nickel and have your student give the amount in cents. Add another nickel, and ask for the total. Continue through 20 nickels, getting your student to count by 5's to 100: 5, 10, 15, 20, 25, 30, 35, 40, 45, 50, 55, 60, 65, 70, 75, 80, 85, 90, 95, 100.

Remind your student that 100 cents is a dollar, so 20 nickels is a dollar.

Set out 1 quarter and have your student give the amount in cents. Put down 5 nickels, one at a time, and have your student count by 5's to 25. Tell your student that a quarter is the same value as 5 nickels.

Set out a quarter. Add another quarter and ask for the total. Repeat with 3 and 4 quarters. Get your student to practice counting by 25's up to 100. Ask your student how many quarters is a dollar.

Discussion

Page 110

Task 1-6, pp. 111-114

Workbook

Exercise 1, p. 186

Reinforcement

As your student goes through these lessons, continue to have him practice counting by fives.

(2) Count coins

Activity

Show a set of coins, less than a dollar. Have several of each type of coin in the set. Have your student count the amount in each set, counting the quarters first, and then counting on by 10's for the dimes, then by 5's for the nickel.

Ask your student if we can count the coins in other ways. Usually it is easier to count the largest value coins first, but sometimes it is helpful to make 30 cents with a quarter and a dime, if we have that. Point out the symbol for cents, ¢. This is written after the amount, e.g. 95¢.

Give your student a handful of coins and ask him to find the total amount of money. Discuss different ways to count the coins, if another way would make sense. Repeat with other sets of coins.

Discussion

Task 7 p. 115

Point out the symbol for cents. Tell your student that we write it after the number.

Task 9, p. 116

Have your student write down the amounts, using the symbol for cents.

Workbook

Exercise 2, pp. 187-188

Reinforcement

Give your students sets of coins totaling less than a dollar and ask him to find the amount of money.

Textbook

Task 7, p. 115

7. 95

Task 9, p. 116

9. (a) 65¢   (b) 57¢
    (c) 80¢   (d) 76¢
    (e) 43¢   (f) 100¢

Workbook

Exercise 2, pp. 187-188

1. 85¢
    35¢
    $1
    25¢
    65¢
2. 80   95
    43   50
    87   80
    73   40

(3) Count sets of bills, make up a set of coins or bills

Textbook

Task 8, p. 115

8. 17

Task 10, p. 117

10. (a) $15.00
     (b) $8.00
     (c) $22.00
     (d) $26.00
     (e) $42.00

Workbook

Exercise 3, pp. 189-190

1. $52
    $8
    $45
    $32
2. (a) 1 quarter, 1 dime,
        1 nickel
    (b) 1 half dollar,
        1 nickels, 1 quarter
    (c) three $5, two $1
    (d) two $10, one $5,
        three $1

Activity

Set out some bills. Have several $20, $10, $5, and $1 bills. Have your student count the amount by starting with the largest bill. Since we do not have $25 bills, it is usually easy to count bills just by starting with the largest type.

Discussion

Task 8, p. 115

Point out the symbol for a dollar, $. This symbol is written in front of the amount, e.g. $17.

Task 10, p. 117

Have your student write down the amounts, using the symbol for the dollar correctly.

Workbook

Exercise 3, pp. 189-190

Enrichment

Show your student a quarter and ask her to find different ways to make a quarter using other coins (5 nickels, or 2 dimes and 1 nickel, or 1 dime, 2 nickels, 5 pennies, etc. As she does this, you can make a chart to keep track of all the different ways that she comes up with. There are 13 different coin combinations for 25 cents.

quarter	dime	nickel	penny
1			
	2	1	
	2		5
	1	3	
	1	2	5
	1	1	10
	1		15
		5	
		4	5
		3	10
		2	15
		1	20
			25

(4) Compare sets of money

Activity

Display two sets of coins. Ask your student to count and compare the amounts of money. Make sure he realizes that a set that has more coins does not necessarily have more money.

Display two sets of bills. Ask your student to count and compare the amounts of money. She should realize that more bills do not necessarily mean more money.

Discussion

Tasks 11-12, p. 118

Workbook

Exercise 4, pp. 191-192

Reinforcement

Have your student compare three sets or more of money.

Enrichment

Display two sets of coins. Discuss ways to compare the amounts of money without counting the amount in each set. For example, look at sets A and B at the top of p. 93. They both have two quarters and 2 dimes, but set A also has a dime and a nickel, and so has more money.

Extra Practice, Unit 19, Exercise 1, pp. 197-200

Tests

Tests, Unit 19, 1A and 1B, pp. 163-170

Textbook

Tasks 11-2, p. 118

5. (a) Set A
   (b) Set X
6. (a) pen
   (b) basketball

Workbook

Exercise 4, pp. 191-192

1. (a) right
   (b) right
2. (a) left
   (b) right
3. top left
4. bottom

Chapter 2 – Shopping

Objectives

♦ Add or subtract amounts of money in cents or dollars.
♦

Material

♦ Items tagged with amounts in cents or dollars.
♦ Coins
♦ Bills

Notes

In this chapter students will use shopping situations to calculate the amount of money needed for buying items. They will use subtraction to find out how much more money they need to buy an item. This helps the student develop number sense and problem-solving abilities.

Set up a play shop by selecting a variety of items and tagging them. Use amounts only in cents or in dollars, not both.

(1) Add or subtract amounts of money in cents or dollars

Activity

Show your student some items tagged with amounts in cents. Give her a dollar bill and tell her she can buy any two items. She needs to pick out two items that are less than $1. Ask her for the total amount the two items cost, and write the addition expression. For example, the two items cost 45¢ and 15¢. Write: 45¢ + 15¢ = 60¢. Ask her to find the answer and then write: 45¢ + 15¢ = 60¢. Then, ask her how much money she will have after she buys the two items. Write the equation, e.g.: 100¢ − 60¢ = 40¢.

Give your student some coins that total less than $1. Pick out an item that is more than the amount of money you gave him. Ask him if he has enough money. Then ask him how much more money he needs. Write the appropriate subtraction equation.

Repeat with a different amount of coins, this time picking out *two* items that together cost either more or less than the amount your student has. Get your student to determine if she has enough money, and if so what she will have left, and if not, how much more she needs.

Discussion

Page 119, Tasks 1-4, pp. 1210-121

For tasks 3 and 4, help your student decide what type of equation is needed. You can use a part-whole model. For example, in task 3 there are two parts, the cost of the bag and the money left over after buying the bag. So to find the amount he had at first, we add the two parts. Write the equations, or ask your student to write them.

Workbook

Exercise 5, pp. 193-195

Reinforcement

Extra Practice, Unit 19, Exercise 2, pp. 201-202

Tests

Tests, Unit 19, 2A and 2B, pp. 171-176

Enrichment

Pick out an item tagged with the cost in cents. Tell your student that he is the cashier. Give him some coins for his cash register. Buy the item using a dollar bill. Guide your student in making change – that is, counting up from the cost of the item using the coins available. Repeat with other items, or several items that add to less than $1.

Textbook

Page 119

Tasks 1-4, pp. 120-121
1. 70; 70
2. 3; 3
3. $18 + $2 = $20
4. 100¢ − 65¢ = 35¢
   Pencil A

Workbook

Exercise 5, pp. 193-195

1. (a) no
   (b) no
   (c) yes
   (d) yes
2. 7
3. 3
4. (a) 15
   (b) 15
   (c) 80
   (d) ice-cream and cake

Review

Review

Review

Workbook

Reviews 14-15, pp. 196-208

Use the review in the workbook as an assessment to see if you need to re-teach any concepts.

If your student has trouble with any of the word problems, get her to act them out, or draw number bonds to help her determine what equation to use.

Tests

Tests, Unit 19 Cumulative Tests A and B, pp. 177-185

Workbook

Review 14, pp. 196-202

1. milk   comics
   pen   apple
2. (a) 72   (b) 85   (c) 40
3. (a) 13 − 9 = 4; The tea set is $4 cheaper than the board game.
   (b) 15 − 12 = 3; 3
   (c) 5 + 10 = 15; 15
4. 10 − 3 = 7; 7
   Mr. Garcia bought 7 more mangoes than Mr. D'cruz; 7
5. 13 + 7 = 20; 20
6. 5 + 6 = 11
7. 12 − 4 = 8
8. 11 − 2 = 9
9. 2, 4, 6, 8, 10, 12, 14, 16
   16
10. 5, 10, 15, 20, 25, 30, 35
11. • → about half past 8
    • → almost half past 7
    • → almost 11 o'clock
    • → almost half past 4
    • → about 7 o'clock
12. (a) >   (b) <   (c) <   (d) <   (e) >
13. bedtime story

Workbook

Review 15, pp. 203-208

1. 70; 80; 78; 98; 68; 48; 50; 80; 79; 99; 89; 90; 100
2. (a) 63   (b) 84
   (c) 57   (d) 78
3. color first, second, and fourth
4. check drawing
5. (a) 6   (b) 2   (c) 2   (d) 18
6. (a) 14   (b) 2
   (c) 8   (d) shirt; guitar
7. $13 - 6 = 7$; 7
8. $20 - 6 = 14$; 14
9. $11 + 3 = 14$; 14
10. $32 + 23 = 55$; 55
11. 45
12. 4
13. David
14. 80
15. $54 - 6 = 48$

Mental Math 1	Mental Math 2	Mental Math 3		Mental Math 4	Mental Math 5	Mental Math 6
1 + 2 = **3**	1 + 1 = **2**	3 + 4 = **7**		1 − 1 = **0**	5 − 4 = **1**	10 − 6 = **4**
5 + 1 = **6**	3 + 6 = **9**	2 + 5 = **7**		5 − 1 = **4**	2 − 2 = **0**	3 − 3 = **0**
9 + 0 = **9**	8 + 2 = **10**	1 + 4 = **5**		3 − 2 = **1**	7 − 3 = **4**	7 − 6 = **1**
7 + 2 = **9**	3 + 7 = **10**	6 + 3 = **9**		9 − 1 = **8**	5 − 5 = **0**	7 − 4 = **3**
2 + 3 = **5**	1 + 7 = **8**	2 + 2 = **4**		8 − 6 = **2**	10 − 3 = **7**	9 − 8 = **1**
4 + 3 = **7**	3 + 2 = **5**	1 + 5 = **6**		8 − 0 = **8**	6 − 1 = **5**	4 − 4 = **0**
7 + 3 = **10**	0 + 4 = **4**	8 + 2 = **10**		9 − 4 = **5**	8 − 4 = **4**	9 − 7 = **2**
2 + 8 = **10**	1 + 6 = **7**	6 + 4 = **10**		6 − 3 = **3**	3 − 1 = **2**	8 − 2 = **6**
5 + 2 = **7**	5 + 5 = **10**	2 + 1 = **3**		8 − 7 = **1**	10 − 4 = **6**	5 − 5 = **0**
4 + 2 = **6**	1 + 8 = **9**	9 + 1 = **10**		10 − 7 = **3**	6 − 2 = **4**	8 − 0 = **8**
4 + 1 = **5**	3 + 7 = **10**	6 + 1 = **7**		8 − 3 = **5**	5 − 3 = **2**	10 − 1 = **9**
4 + 4 = **8**	5 + 5 = **10**	7 + 0 = **7**		9 − 6 = **3**	10 − 8 = **2**	4 − 3 = **1**
2 + 6 = **8**	3 + 3 = **6**	4 + 5 = **9**		2 − 1 = **1**	7 − 2 = **5**	2 − 1 = **1**
8 + 1 = **9**	2 + 7 = **9**	6 + 4 = **10**		10 − 10 = **0**	5 − 0 = **5**	8 − 5 = **3**
6 + 4 = **10**	4 + 6 = **10**	3 + 5 = **8**		6 − 4 = **2**	4 − 2 = **2**	10 − 9 = **1**
3 + 1 = **4**	5 + 3 = **8**	1 + 3 = **4**		9 − 9 = **0**	8 − 1 = **7**	9 − 3 = **6**
0 + 8 = **8**	0 + 9 = **9**	2 + 4 = **6**		7 − 1 = **6**	9 − 5 = **4**	7 − 7 = **0**
7 + 1 = **8**	5 + 4 = **9**	6 + 2 = **8**		10 − 5 = **5**	9 − 2 = **7**	5 − 2 = **3**
1 + 9 = **10**	1 + 9 = **10**	1 + 0 = **1**		7 − 5 = **2**	10 − 2 = **8**	6 − 5 = **1**
3 + 0 = **3**	7 + 3 = **10**	3 + 7 = **10**		4 − 1 = **3**	6 − 6 = **0**	8 − 8 = **0**

Mental Math 7	Mental Math 8	Mental Math 9		Mental Math 10	Mental Math 11	Mental Math 12
5 + 1 = **6**	21 + 2 = **23**	27 + 1 = **28**		7 + 4 = **11**	7 + 8 = **15**	3 + 7 = **10**
3 − 2 = **1**	29 − 3 = **26**	29 + 2 = **31**		9 + 3 = **12**	6 + 5 = **11**	5 + 7 = **12**
2 + 3 = **5**	32 + 4 = **36**	36 − 3 = **33**		6 + 6 = **12**	7 + 9 = **16**	6 + 7 = **13**
9 − 4 = **5**	13 + 2 = **15**	28 − 3 = **25**		8 + 5 = **13**	5 + 7 = **12**	5 + 1 = **6**
10 − 7 = **3**	38 − 5 = **33**	28 + 3 = **31**		8 + 6 = **14**	6 + 7 = **13**	6 + 8 = **14**
4 + 3 = **7**	27 − 4 = **23**	30 − 2 = **28**		8 + 9 = **17**	7 + 8 = **15**	2 + 7 = **9**
2 + 7 = **9**	34 + 3 = **37**	29 + 1 = **30**		9 + 7 = **16**	6 + 8 = **14**	4 + 7 = **11**
2 − 1 = **1**	36 − 5 = **31**	22 + 2 = **24**		9 + 6 = **15**	4 + 7 = **11**	9 + 6 = **15**
7 + 2 = **9**	15 + 5 = **20**	37 − 3 = **34**		7 + 7 = **14**	7 + 3 = **10**	1 + 8 = **9**
9 − 9 = **0**	26 + 4 = **30**	31 − 3 = **28**		9 + 4 = **13**	6 + 9 = **15**	8 + 5 = **13**
7 − 5 = **2**	16 + 3 = **19**	23 + 2 = **25**		7 + 5 = **12**	7 + 9 = **16**	7 + 6 = **13**
2 + 5 = **7**	25 − 2 = **23**	38 − 1 = **37**		6 + 4 = **10**	3 + 8 = **11**	3 + 3 = **6**
2 − 2 = **0**	34 − 3 = **31**	29 + 1 = **30**		8 + 3 = **11**	4 + 8 = **12**	8 + 3 = **11**
5 + 3 = **8**	37 + 1 = **38**	19 + 2 = **21**		9 + 9 = **18**	5 + 6 = **11**	2 + 7 = **9**
6 − 1 = **5**	29 − 6 = **23**	38 − 3 = **35**		9 + 8 = **17**	3 + 9 = **12**	9 + 9 = **18**
10 − 4 = **6**	28 + 2 = **30**	31 − 2 = **29**		8 + 8 = **16**	4 + 9 = **13**	4 + 6 = **10**
7 + 3 = **10**	17 − 7 = **10**	39 + 1 = **40**		8 + 7 = **15**	5 + 8 = **13**	5 + 6 = **11**
3 + 5 = **8**	20 + 9 = **29**	27 − 3 = **24**		9 + 5 = **14**	5 + 9 = **14**	5 + 3 = **8**
9 + 1 = **10**	38 − 3 = **35**	19 + 3 = **22**		7 + 6 = **13**	2 + 9 = **11**	7 + 6 = **13**
7 − 2 = **5**	22 − 2 = **20**	27 + 2 = **29**		8 + 4 = **12**	8 + 8 = **16**	8 + 4 = **12**

Mental Math 13	Mental Math 14	Mental Math 15		Mental Math 16	Mental Math 17	Mental Math 18
27 + 4 = **31**	11 − 2 = **9**	10 − 3 = **7**		11 − 3 = **8**	30 − 5 = **25**	65 + 2 = **67**
19 + 3 = **22**	12 − 5 = **7**	20 − 5 = **15**		22 − 6 = **16**	21 − 8 = **13**	45 + 5 = **50**
26 + 6 = **32**	13 − 4 = **9**	30 − 4 = **26**		35 − 7 = **28**	26 + 9 = **35**	85 + 6 = **91**
27 + 9 = **36**	14 − 6 = **8**	40 − 9 = **31**		43 − 9 = **34**	32 + 8 = **40**	29 + 9 = **38**
18 + 5 = **23**	15 − 9 = **6**	10 − 7 = **3**		26 − 7 = **19**	22 − 3 = **19**	73 + 5 = **78**
24 + 7 = **31**	16 − 8 = **8**	20 − 6 = **14**		22 − 7 = **15**	40 − 7 = **33**	42 + 8 = **50**
18 + 6 = **24**	17 − 8 = **9**	30 − 1 = **29**		31 − 2 = **29**	26 + 4 = **30**	39 + 4 = **43**
21 + 9 = **30**	18 − 7 = **11**	40 − 8 = **32**		36 − 8 = **28**	22 + 9 = **31**	95 + 2 = **97**
32 + 8 = **40**	18 − 9 = **9**	30 − 3 = **27**		23 − 10 = **13**	24 − 6 = **18**	88 + 5 = **93**
18 + 9 = **27**	11 − 4 = **7**	20 − 10 = **10**		37 − 2 = **35**	31 − 7 = **24**	58 + 2 = **60**
29 + 7 = **36**	12 − 3 = **9**	40 − 2 = **38**		36 − 7 = **29**	23 + 8 = **31**	74 + 7 = **81**
28 + 3 = **31**	13 − 5 = **8**	30 − 20 = **10**		13 − 5 = **8**	33 + 4 = **37**	90 + 7 = **97**
19 + 6 = **25**	14 − 7 = **7**	19 + 6 = **25**		32 − 7 = **25**	12 − 9 = **3**	59 + 1 = **60**
25 + 8 = **33**	15 − 6 = **9**	32 + 7 = **39**		35 − 9 = **26**	15 + 8 = **23**	68 + 8 = **76**
33 + 6 = **39**	16 − 9 = **7**	12 − 5 = **7**		27 − 10 = **17**	37 − 5 = **32**	73 + 6 = **79**
26 + 8 = **34**	17 − 9 = **8**	8 + 4 = **12**		28 − 5 = **23**	25 + 3 = **28**	68 + 3 = **71**
27 + 6 = **33**	11 − 8 = **3**	32 + 8 = **40**		28 − 9 = **19**	33 − 7 = **26**	87 + 7 = **94**
18 + 8 = **26**	12 − 6 = **6**	30 − 7 = **23**		32 − 4 = **28**	27 − 5 = **22**	49 + 6 = **55**
31 + 9 = **40**	13 − 7 = **6**	29 + 6 = **35**		27 − 6 = **21**	21 − 9 = **12**	59 + 2 = **61**
25 + 7 = **32**	14 − 5 = **9**	30 − 9 = **21**		40 − 6 = **34**	21 + 9 = **30**	54 + 8 = **62**

Mental Math 19	Mental Math 20	Mental Math 21		Mental Math 22	Mental Math 23	Mental Math 24
73 + 6 = **79**	70 + 20 = **90**	13 + 15 = **28**		18 + 12 = **30**	89 − 1 = **88**	81 − 6 = **75**
93 + 7 = **100**	20 + 60 = **80**	64 + 15 = **79**		18 + 15 = **33**	81 − 9 = **72**	53 − 6 = **47**
63 + 8 = **71**	50 + 20 = **70**	71 + 23 = **94**		28 + 35 = **63**	93 − 5 = **88**	74 − 6 = **68**
47 + 5 = **52**	28 + 40 = **68**	14 + 42 = **56**		14 + 47 = **61**	32 − 3 = **29**	48 − 9 = **39**
59 + 5 = **64**	80 + 17 = **97**	71 + 15 = **86**		79 + 15 = **94**	54 − 9 = **45**	38 − 2 = **36**
1 + 94 = **95**	32 + 20 = **52**	41 + 25 = **66**		48 + 25 = **73**	63 − 8 = **55**	67 − 7 = **60**
87 + 3 = **90**	30 + 36 = **66**	16 + 72 = **88**		16 + 74 = **90**	75 − 7 = **68**	11 − 7 = **4**
6 + 66 = **72**	40 + 33 = **73**	13 + 82 = **95**		23 + 62 = **85**	26 − 8 = **18**	27 − 9 = **18**
64 + 6 = **70**	35 + 50 = **85**	21 + 21 = **42**		28 + 28 = **56**	47 − 4 = **43**	72 − 5 = **67**
5 + 54 = **59**	64 + 30 = **94**	25 + 31 = **56**		25 + 37 = **62**	15 − 6 = **9**	53 − 7 = **46**
8 + 49 = **57**	40 + 43 = **83**	52 + 44 = **96**		55 + 36 = **91**	83 − 4 = **79**	85 − 5 = **80**
77 + 8 = **85**	22 + 60 = **82**	62 + 35 = **97**		42 + 35 = **77**	67 − 8 = **59**	36 − 3 = **33**
2 + 46 = **48**	40 + 54 = **94**	42 + 16 = **58**		42 + 19 = **61**	73 − 9 = **64**	42 − 4 = **38**
78 + 7 = **85**	15 + 70 = **85**	81 + 12 = **93**		37 + 12 = **49**	42 − 7 = **35**	92 − 9 = **83**
3 + 84 = **87**	60 + 16 = **76**	42 + 37 = **79**		48 + 37 = **85**	94 − 7 = **87**	21 − 5 = **16**
58 + 6 = **64**	50 + 12 = **62**	33 + 31 = **64**		33 + 39 = **72**	92 − 6 = **86**	84 − 5 = **79**
37 + 6 = **43**	17 + 40 = **57**	35 + 42 = **77**		38 + 42 = **80**	56 − 7 = **49**	11 − 4 = **7**
9 + 43 = **52**	39 + 10 = **49**	23 + 15 = **38**		26 + 16 = **42**	48 − 7 = **41**	81 − 3 = **78**
80 + 5 = **85**	58 + 40 = **98**	83 + 16 = **99**		84 + 16 = **100**	31 − 8 = **23**	61 − 2 = **59**
92 + 7 = **99**	50 + 31 = **81**	45 + 43 = **88**		45 + 45 = **90**	16 − 9 = **7**	52 − 8 = **44**

Mental Math 25	Mental Math 26	Mental Math 27
80 – 10 = **70**	79 – 71 = **8**	40 – 3 = **37**
70 – 20 = **50**	69 – 12 = **57**	40 + 3 = **43**
90 – 30 = **60**	78 – 46 = **32**	85 – 8 = **77**
60 – 40 = **20**	57 – 11 = **46**	85 – 28 = **57**
68 – 40 = **28**	96 – 42 = **54**	73 + 27 = **100**
47 – 30 = **17**	59 – 25 = **34**	38 + 4 = **42**
92 – 60 = **32**	69 – 36 = **33**	42 + 5 = **47**
86 – 40 = **46**	59 – 57 = **2**	85 – 14 = **71**
57 – 20 = **37**	69 – 48 = **21**	22 + 9 = **31**
81 – 50 = **31**	93 – 91 = **2**	6 + 48 = **54**
74 – 30 = **44**	87 – 17 = **70**	92 – 24 = **68**
98 – 80 = **18**	58 – 32 = **26**	42 + 38 = **80**
79 – 30 = **49**	49 – 13 = **36**	2 + 81 = **83**
79 – 3 = **76**	85 – 37 = **48**	17 – 9 = **8**
83 – 60 = **23**	82 – 44 = **38**	36 + 48 = **84**
83 – 6 = **77**	65 – 58 = **7**	46 – 22 = **24**
64 – 8 = **56**	85 – 74 = **11**	92 – 15 = **77**
68 – 40 = **28**	48 – 38 = **10**	47 – 23 = **24**
25 – 9 = **16**	76 – 26 = **50**	42 – 18 = **24**
65 – 8 = **57**	33 – 27 = **6**	30 + 42 = **72**

Mental Math 1	Mental Math 2	Mental Math 3
1 + 2 = _____	1 + 1 = _____	3 + 4 = _____
5 + 1 = _____	3 + 6 = _____	2 + 5 = _____
9 + 0 = _____	8 + 2 = _____	1 + 4 = _____
7 + 2 = _____	3 + 7 = _____	6 + 3 = _____
2 + 3 = _____	1 + 7 = _____	2 + 2 = _____
4 + 3 = _____	3 + 2 = _____	1 + 5 = _____
7 + 3 = _____	0 + 4 = _____	8 + 2 = _____
2 + 8 = _____	1 + 6 = _____	6 + 4 = _____
5 + 2 = _____	5 + 5 = _____	2 + 1 = _____
4 + 2 = _____	1 + 8 = _____	9 + 1 = _____
4 + 1 = _____	3 + 7 = _____	6 + 1 = _____
4 + 4 = _____	5 + 5 = _____	7 + 0 = _____
2 + 6 = _____	3 + 3 = _____	4 + 5 = _____
8 + 1 = _____	2 + 7 = _____	6 + 4 = _____
6 + 4 = _____	4 + 6 = _____	3 + 5 = _____
3 + 1 = _____	5 + 3 = _____	1 + 3 = _____
0 + 8 = _____	0 + 9 = _____	2 + 4 = _____
7 + 1 = _____	5 + 4 = _____	6 + 2 = _____
1 + 9 = _____	1 + 9 = _____	1 + 0 = _____
3 + 0 = _____	7 + 3 = _____	3 + 7 = _____

Mental Math 4	Mental Math 5	Mental Math 6
$1 - 1 =$ _____	$5 - 4 =$ _____	$10 - 6 =$ _____
$5 - 1 =$ _____	$2 - 2 =$ _____	$3 - 3 =$ _____
$3 - 2 =$ _____	$7 - 3 =$ _____	$7 - 6 =$ _____
$9 - 1 =$ _____	$5 - 5 =$ _____	$7 - 4 =$ _____
$8 - 6 =$ _____	$10 - 3 =$ _____	$9 - 8 =$ _____
$8 - 0 =$ _____	$6 - 1 =$ _____	$4 - 4 =$ _____
$9 - 4 =$ _____	$8 - 4 =$ _____	$9 - 7 =$ _____
$6 - 3 =$ _____	$3 - 1 =$ _____	$8 - 2 =$ _____
$8 - 7 =$ _____	$10 - 4 =$ _____	$5 - 5 =$ _____
$10 - 7 =$ _____	$6 - 2 =$ _____	$8 - 0 =$ _____
$8 - 3 =$ _____	$5 - 3 =$ _____	$10 - 1 =$ _____
$9 - 6 =$ _____	$10 - 8 =$ _____	$4 - 3 =$ _____
$2 - 1 =$ _____	$7 - 2 =$ _____	$2 - 1 =$ _____
$10 - 10 =$ _____	$5 - 0 =$ _____	$8 - 5 =$ _____
$6 - 4 =$ _____	$4 - 2 =$ _____	$10 - 9 =$ _____
$9 - 9 =$ _____	$8 - 1 =$ _____	$9 - 3 =$ _____
$7 - 1 =$ _____	$9 - 5 =$ _____	$7 - 7 =$ _____
$10 - 5 =$ _____	$9 - 2 =$ _____	$5 - 2 =$ _____
$7 - 5 =$ _____	$10 - 2 =$ _____	$6 - 5 =$ _____
$4 - 1 =$ _____	$6 - 6 =$ _____	$8 - 8 =$ _____

Mental Math 7	Mental Math 8	Mental Math 9
5 + 1 = _____	21 + 2 = _____	27 + 1 = _____
3 − 2 = _____	29 − 3 = _____	29 + 2 = _____
2 + 3 = _____	32 + 4 = _____	36 − 3 = _____
9 − 4 = _____	13 + 2 = _____	28 − 3 = _____
10 − 7 = _____	38 − 5 = _____	28 + 3 = _____
4 + 3 = _____	27 − 4 = _____	30 − 2 = _____
2 + 7 = _____	34 + 3 = _____	29 + 1 = _____
2 − 1 = _____	36 − 5 = _____	22 + 2 = _____
7 + 2 = _____	15 + 5 = _____	37 − 3 = _____
9 − 9 = _____	26 + 4 = _____	31 − 3 = _____
7 − 5 = _____	16 + 3 = _____	23 + 2 = _____
2 + 5 = _____	25 − 2 = _____	38 − 1 = _____
2 − 2 = _____	34 − 3 = _____	29 + 1 = _____
5 + 3 = _____	37 + 1 = _____	19 + 2 = _____
6 − 1 = _____	29 − 6 = _____	38 − 3 = _____
10 − 4 = _____	28 + 2 = _____	31 − 2 = _____
7 + 3 = _____	17 − 7 = _____	39 + 1 = _____
3 + 5 = _____	20 + 9 = _____	27 − 3 = _____
9 + 1 = _____	38 − 3 = _____	19 + 2 = _____
7 − 2 = _____	22 − 2 = _____	27 + 2 = _____

Mental Math 10	Mental Math 11	Mental Math 12
7 + 4 = _____	7 + 8 = _____	3 + 7 = _____
9 + 3 = _____	6 + 5 = _____	5 + 7 = _____
6 + 6 = _____	7 + 9 = _____	6 + 7 = _____
8 + 5 = _____	5 + 7 = _____	5 + 1 = _____
8 + 6 = _____	6 + 7 = _____	6 + 8 = _____
8 + 9 = _____	7 + 8 = _____	2 + 7 = _____
9 + 7 = _____	6 + 8 = _____	4 + 7 = _____
9 + 6 = _____	4 + 7 = _____	9 + 6 = _____
7 + 7 = _____	7 + 3 = _____	1 + 8 = _____
9 + 4 = _____	6 + 9 = _____	8 + 5 = _____
7 + 5 = _____	7 + 9 = _____	7 + 6 = _____
6 + 4 = _____	3 + 8 = _____	3 + 3 = _____
8 + 3 = _____	4 + 8 = _____	8 + 3 = _____
9 + 9 = _____	5 + 6 = _____	2 + 7 = _____
9 + 8 = _____	3 + 9 = _____	9 + 9 = _____
8 + 8 = _____	4 + 9 = _____	4 + 6 = _____
8 + 7 = _____	5 + 8 = _____	5 + 6 = _____
9 + 5 = _____	5 + 9 = _____	5 + 3 = _____
7 + 6 = _____	2 + 9 = _____	7 + 6 = _____
8 + 4 = _____	8 + 8 = _____	8 + 4 = _____

Mental Math 13	Mental Math 14	Mental Math 15
27 + 4 = _____	11 – 2 = _____	10 – 3 = _____
19 + 3 = _____	12 – 5 = _____	20 – 5 = _____
26 + 6 = _____	13 – 4 = _____	30 – 4 = _____
27 + 9 = _____	14 – 6 = _____	40 – 9 = _____
18 + 5 = _____	15 – 9 = _____	10 – 7 = _____
24 + 7 = _____	16 – 8 = _____	20 – 6 = _____
18 + 6 = _____	17 – 8 = _____	30 – 1 = _____
21 + 9 = _____	18 – 7 = _____	40 – 8 = _____
32 + 8 = _____	18 – 9 = _____	30 – 3 = _____
18 + 9 = _____	11 – 4 = _____	20 – 10 = _____
29 + 7 = _____	12 – 3 = _____	40 – 2 = _____
28 + 3 = _____	13 – 5 = _____	30 – 20 = _____
19 + 6 = _____	14 – 7 = _____	19 + 6 = _____
25 + 8 = _____	15 – 6 = _____	32 + 7 = _____
33 + 6 = _____	16 – 9 = _____	12 – 5 = _____
26 + 8 = _____	17 – 9 = _____	8 + 4 = _____
27 + 6 = _____	11 – 8 = _____	32 + 8 = _____
18 + 8 = _____	12 – 6 = _____	30 – 7 = _____
31 + 9 = _____	13 – 7 = _____	29 + 6 = _____
25 + 7 = _____	14 – 5 = _____	30 – 9 = _____

Mental Math 16	Mental Math 17	Mental Math 18
11 − 3 = _____	30 − 5 = _____	65 + 2 = _____
22 − 6 = _____	21 − 8 = _____	45 + 5 = _____
35 − 7 = _____	26 + 9 = _____	85 + 6 = _____
43 − 9 = _____	32 + 8 = _____	29 + 9 = _____
26 − 7 = _____	22 − 3 = _____	73 + 5 = _____
22 − 7 = _____	40 − 7 = _____	42 + 8 = _____
31 − 2 = _____	26 + 4 = _____	39 + 4 = _____
36 − 8 = _____	22 + 9 = _____	95 + 2 = _____
23 − 10 = _____	24 − 6 = _____	88 + 5 = _____
37 − 2 = _____	31 − 7 = _____	58 + 2 = _____
36 − 7 = _____	23 + 8 = _____	74 + 7 = _____
13 − 5 = _____	33 + 4 = _____	90 + 7 = _____
32 − 7 = _____	12 − 9 = _____	59 + 1 = _____
35 − 9 = _____	15 + 8 = _____	68 + 8 = _____
27 − 10 = _____	37 − 5 = _____	73 + 6 = _____
28 − 5 = _____	25 + 3 = _____	68 + 3 = _____
28 − 9 = _____	33 − 7 = _____	87 + 7 = _____
32 − 4 = _____	27 − 5 = _____	49 + 6 = _____
27 − 6 = _____	21 − 9 = _____	59 + 2 = _____
40 − 6 = _____	21 + 9 = _____	54 + 8 = _____

Mental Math 19	Mental Math 20	Mental Math 21
73 + 6 = _____	70 + 20 = _____	13 + 15 = _____
93 + 7 = _____	20 + 60 = _____	64 + 15 = _____
63 + 8 = _____	50 + 20 = _____	71 + 23 = _____
47 + 5 = _____	28 + 40 = _____	14 + 42 = _____
59 + 5 = _____	80 + 17 = _____	71 + 15 = _____
1 + 94 = _____	32 + 20 = _____	41 + 25 = _____
87 + 3 = _____	30 + 36 = _____	16 + 72 = _____
6 + 66 = _____	40 + 33 = _____	13 + 82 = _____
64 + 6 = _____	35 + 50 = _____	21 + 21 = _____
5 + 54 = _____	64 + 30 = _____	25 + 31 = _____
8 + 49 = _____	40 + 43 = _____	52 + 44 = _____
77 + 8 = _____	22 + 60 = _____	62 + 35 = _____
2 + 46 = _____	40 + 54 = _____	42 + 16 = _____
78 + 7 = _____	15 + 70 = _____	81 + 12 = _____
3 + 84 = _____	60 + 16 = _____	42 + 37 = _____
58 + 6 = _____	50 + 12 = _____	33 + 31 = _____
37 + 6 = _____	17 + 40 = _____	35 + 42 = _____
9 + 43 = _____	39 + 10 = _____	23 + 15 = _____
80 + 5 = _____	58 + 40 = _____	83 + 16 = _____
92 + 7 = _____	50 + 31 = _____	45 + 43 = _____

Mental Math 22	Mental Math 23	Mental Math 24
18 + 12 = _____	89 − 1 = _____	81 − 6 = _____
18 + 15 = _____	81 − 9 = _____	53 − 6 = _____
28 + 35 = _____	93 − 5 = _____	74 − 6 = _____
14 + 47 = _____	32 − 3 = _____	48 − 9 = _____
79 + 15 = _____	54 − 9 = _____	38 − 2 = _____
48 + 25 = _____	63 − 8 = _____	67 − 7 = _____
16 + 74 = _____	75 − 7 = _____	11 − 7 = _____
23 + 62 = _____	26 − 8 = _____	27 − 9 = _____
28 + 28 = _____	47 − 4 = _____	72 − 5 = _____
25 + 37 = _____	15 − 6 = _____	53 − 7 = _____
55 + 36 = _____	83 − 4 = _____	85 − 5 = _____
42 + 35 = _____	67 − 8 = _____	36 − 3 = _____
42 + 19 = _____	73 − 9 = _____	42 − 4 = _____
37 + 12 = _____	42 − 7 = _____	92 − 9 = _____
48 + 37 = _____	94 − 7 = _____	21 − 5 = _____
33 + 39 = _____	92 − 6 = _____	84 − 5 = _____
38 + 42 = _____	56 − 7 = _____	11 − 4 = _____
26 + 16 = _____	48 − 7 = _____	81 − 3 = _____
84 + 16 = _____	31 − 8 = _____	61 − 2 = _____
45 + 45 = _____	16 − 9 = _____	52 − 8 = _____

Mental Math 25	Mental Math 26	Mental Math 27
80 – 10 = _____	79 – 71 = _____	40 – 3 = _____
70 – 20 = _____	69 – 12 = _____	40 + 3 = _____
90 – 30 = _____	78 – 46 = _____	85 – 8 = _____
60 – 40 = _____	57 – 11 = _____	85 – 28 = _____
68 – 40 = _____	96 – 42 = _____	73 + 27 = _____
47 – 30 = _____	59 – 25 = _____	38 + 4 = _____
92 – 60 = _____	69 – 36 = _____	42 + 5 = _____
86 – 40 = _____	59 – 57 = _____	85 – 14 = _____
57 – 20 = _____	69 – 48 = _____	22 + 9 = _____
81 – 50 = _____	93 – 91 = _____	6 + 48 = _____
74 – 30 = _____	87 – 17 = _____	92 – 24 = _____
98 – 80 = _____	58 – 32 = _____	42 + 38 = _____
79 – 30 = _____	49 – 13 = _____	2 + 81 = _____
79 – 3 = _____	85 – 37 = _____	17 – 9 = _____
83 – 60 = _____	82 – 44 = _____	36 + 48 = _____
83 – 6 = _____	65 – 58 = _____	46 – 22 = _____
64 – 8 = _____	85 – 74 = _____	92 – 15 = _____
68 – 40 = _____	48 – 38 = _____	47 – 23 = _____
25 – 9 = _____	76 – 26 = _____	42 – 18 = _____
65 – 8 = _____	33 – 27 = _____	30 + 42 = _____

Monkey										
Lion										
Bear										

1	2	3	4	5	6	7	8	9	10
11	12	13	14	15	16	17	18	19	20
21	22	23	24	25	26	27	28	29	30
31	32	33	34	35	36	37	38	39	40

1	2	3	4	5	6	7	8	9	10
11	12	13	14	15	16	17	18	19	20
21	22	23	24	25	26	27	28	29	30
31	32	33	34	35	36	37	38	39	40
41	42	43	44	45	46	47	48	49	50
51	52	53	54	55	56	57	58	59	60
61	62	63	64	65	66	67	68	69	70
71	72	73	74	75	76	77	78	79	80
81	82	83	84	85	86	87	88	89	90
91	92	93	94	95	96	97	98	99	100

A B C D E

F G H I J

K L M N O

P Q R S T

U V W X Y Z

1 0   1

2 0   2

3 0   3

4 0 4

5 0 5

6 0 6

7 0 7

8 0 8

9 0 9

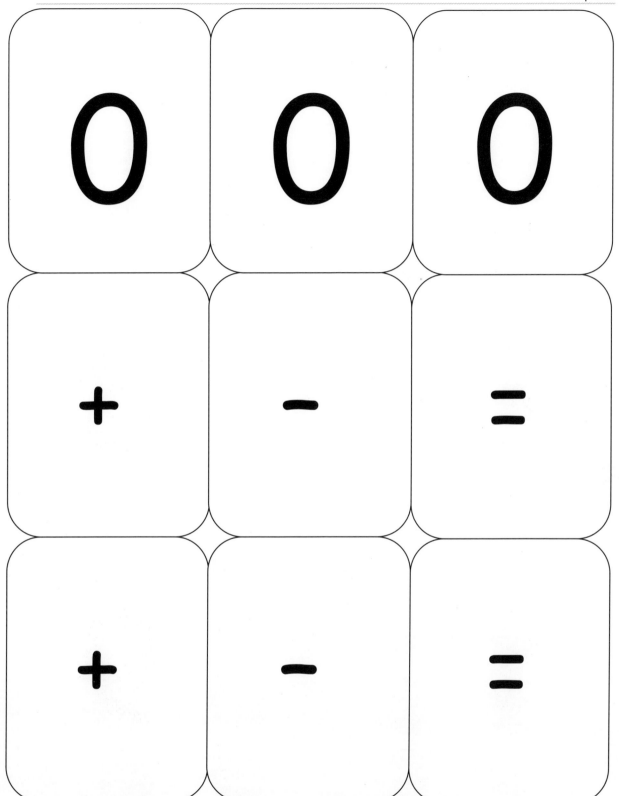

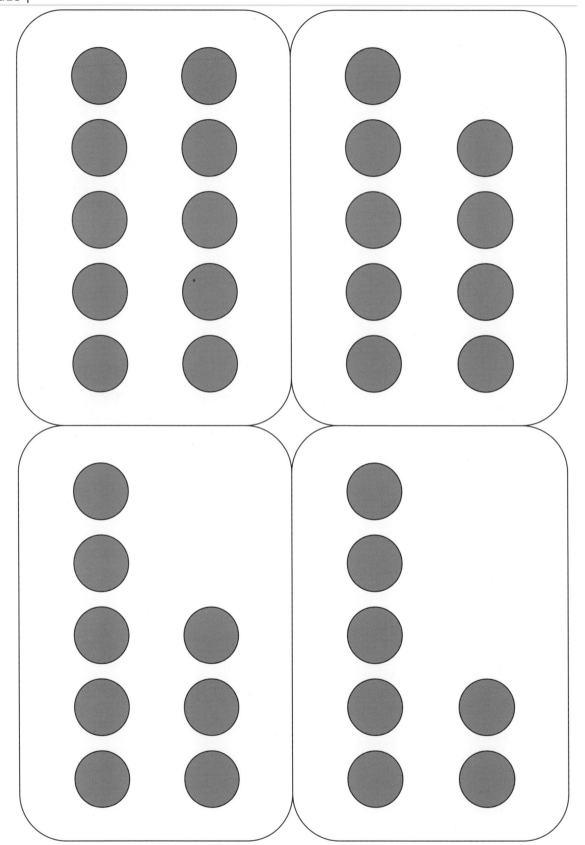

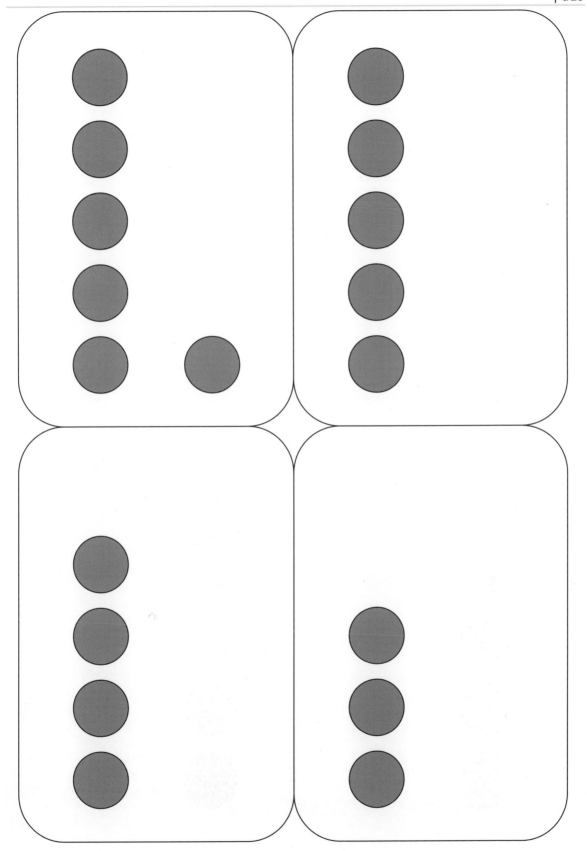

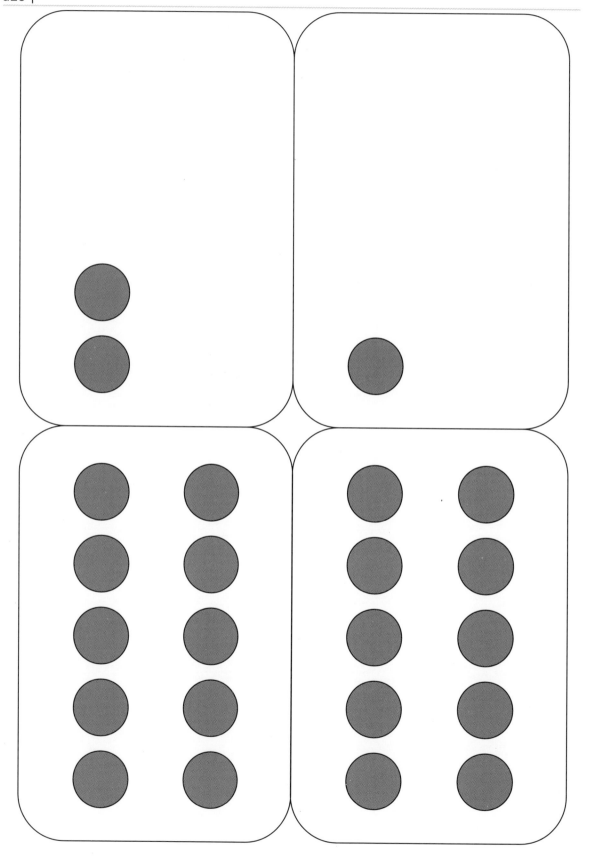